CISM COURSES AND LECTURES

Series Editors:

The Rectors of CISM
Sandor Kaliszky - Budapest
Mahir Sayir - Zurich
Wilhelm Schneider - Wien

The Secretary General of CISM
Giovanni Bianchi - Milan

Executive Editor
Carlo Tasso - Udine

The series presents lecture notes, monographs, edited works and proceedings in the field of Mechanics, Engineering, Computer Science and Applied Mathematics.
Purpose of the series is to make known in the international scientific and technical community results obtained in some of the activities organized by CISM, the International Centre for Mechanical Sciences.

DEVELOPMENT OF KNOWLEDGE-BASED SYSTEMS FOR ENGINEERING

EDITED BY

CARLO TASSO
UNIVERSITY OF UDINE

EDOARDO R. DE ARANTES E OLIVEIRA
NATIONAL LABORATORY OF CIVIL ENGINEERING OF LISBON

Springer-Verlag Wien GmbH

Le spese di stampa di questo volume sono in parte coperte da
contributi del Consiglio Nazionale delle Ricerche.

This volume contains 88 illustrations

In order to make this volume available as economically and as
rapidly as possible the authors' typescripts have been
reproduced in their original forms. This method unfortunately
has its typographical limitations but it is hoped that they in no
way distract the reader.

ISBN 978-3-211-82916-5   ISBN 978-3-7091-2784-1 (eBook)
DOI 10.1007/978-3-7091-2784-1

PREFACE

The application of Artificial Intelligence techniques to specific Engineering problems is a natural extension of the exploitation and widespread development of information technology. Among such applications, expert systems and more generally knowledge-based systems (KBSs) play a fundamental role: these sophisticated programs are aimed at emulating complex human problem solving capabilities by exploiting the explicit representation of expert domain knowledge and by means of sophisticated reasoning algorithms. Knowledge-based systems feature processing capabilities which go beyond the classical numerical applications of computers to Engineering. They have been utilized for performing complex Engineering tasks, such as design, diagnosis, data interpretation, scheduling, planning, monitoring and so on.

The goal of this volume is to help engineers to better understand the design and development process and the specific techniques utilized for constructing expert systems in Engineering.

The first paper by de Arantes e Oliveira and Bento provides a perspective view of structural Engineering design, ranging from basic established principles to modern approaches, including CAD, ICAD, and KBS systems.

The second paper by Tasso gives the fundamental definitions of artificial intelligence and knowledge-based systems. The paper is also aimed at providing an extensive bibliography to basic concepts and techniques.

The paper by Chung discusses some of the proven knowledge elicitation techniques aimed at acquiring from single human experts the relevant domain knowledge to be embodied in a KBS.

The following contribution by Bento illustrates how mechanical behavior can be modeled by means of 'non-physical' tools, namely artificial neural networks, which clearly overwhelm the traditional approaches based on computational or theoretical mechanics.

In their paper, Fenves and Turkiyyah describe and evaluate two approaches to interfacing and integrating knowledge-based systems with numerical finite element processes for applications such as model generation, model interpretation, integration with design, and comprehensive design environments.

The paper by Terk and Fenves deals with integrated building design by illustrating a specific project, providing its motivation and history, as well as an overview of the knowledge-based agents participating in the design process, and a brief extrapolation to the future of integrated design environments.

In the following contribution, Sriram covers the problem of engineering a product by means of computer aided tools specifically aimed at exploiting the collaborative nature of the development process, which can greatly profit from the cooperation of the various engineering disciplines and skills involved.

The paper by Maher describes how machine learning techniques can be exploited for automatically learning design knowledge: more specifically she presents a conceptual clustering program aimed at deriving functional, structural, and behavioral knowledge included in a design prototype.

Feijò explores the possibilities of virtual environments technologies in CAD systems, it clarifies the nature of these tools and provides a practical guide to CAD researchers and engineers.

The following contribution by Toppano is aimed at defining an abstract model of the design process, and specifically deals with conceptual design of technical systems based on the exploitation of multiple representations of the artifact to be designed and on the use of specific patterns of inference, called transmutations, to be exploited in multistrategy design.

The last paper by Chung and Stone provides an overview of the work that has been done for applying computer technologies to regulatory information: they emphasize the use of advanced information technologies for supporting the access, interpretation, and exploitation of information.

We wish to thank all the authors for their valuable contribution to this volume and to all the participants to the CISM advanced school on "Development of Expert Systems for Structural Mechanics and Structural Engineering", for their active participation and for the interesting and fruitful discussions. Special thanks to the CISM Staff for the valuable support.

Carlo Tasso
Edoardo R. de Arantes e Oliveira

CONTENTS

Page

THE SENSE OF PROGRESS IN STRUCTURAL ENGINEERING

E.R. de Arantes e Oliveira
National Laboratory of Civil Engineering, Lisbon, Portugal

J. Bento
Technical University of Lisbon, Lisbon, Portugal

1. Introduction

The Spanish philosopher Ortega y Gasset, referring to Medicine, as he might have been referring to Engineering, recalled that it is not in itself a science, but a profession, i.e. a practical activity. As such, its point of view is distinct from that of a Science : "... goes to Science, takes over everything that seems to satisfy it, takes from its results what it considers most effective, and leaves the rest".

According to Ortega, "... it is there in order to find solutions. If they are scientific, so much the better, but it is not essential that they should be. They may come from a millenary experience that experience neither explains nor underwrites".

That these words are valid for structural engineering is shown by the fact that for so many centuries it has existed without making use of Science, but of a "millenary experience".

The principles on which the art of constructing was based had above all the character of mnemonics, which tended to be received by initiates as elements of a revelation in which only some, those exceptionally gifted, dared to introduce modifications.

Such a situation would be incompatible with a conjuncture like the present one, in which, constructing more, more rapidly and more economically than in any other epoch, not only has it become a permanent need to innovate, but also the profession has come to be exercised by an increasingly numerous body of engineers, among whom the exceptionally gifted cannot be more than a bare minority.

The trend of progress in Structural Engineering is therefore for it to become more transparent and less mysterious, more rational and less intuitive, more scientific and less empirical.

2. MODELLING

Structural engineering could make use of the scientific method without becoming a privileged field of application of mechanics (Arantes e Oliveira and Pedro, 1986). Artificial intelligence, for example, may provide completely novel approaches, as shown latter in the book.

One possible scientific method of forecasting the behaviour of a structure consists in examining the behaviour of an identical structure. But even this method, rudimentary although scientific, presents its difficulties owing to the simple fact that no two perfectly equal objects exist in Nature.

Specifically: stating that two beams of reinforced concrete are equal cannot mean that they are in fact so, but that certain properties of the concrete and steel in both coincide, that their geometry and dimensions are the same and that the reinforcement rods have the same layout and diameters - all of this, of course, within tolerances regarded as admissible. Such a statement thus presupposes a theory which sates that the above mentioned parameters are those that determine the behaviour of the beams. More precisely, a reinforced concrete beam is identified with a set of properties that define it as an ideal object which can be made to correspond, in the real world, not exclusively to the beam concerned, but to an infinite number of beams that have the same properties.

What occurs in structural engineering is not exclusive to it, but is typical of the process through which the human mind knows the real world.

Since the real world, is inaccessible to the mind, we can only know it through ideas, which, contrarily to what was asserted in the Middle Ages by the conceptualists, correspond in fact to elements of the real world. But such elements exist in this world in a singular and individual form, and not in the form by which they exist in the mind.

The universe of knowledge is not, therefore, a copy of the objective universe, but a construction of the intelligence based on sensitive data provided by experience. This is the position which in the Theory of Knowledge, is given the name of *critical* or *moderate realism*. In adopting such position, the modern scientist establishes a correspondence between the beings of the real world and the elements of an ideal world called *idealizations* or *models* of the former. When such models consist of a set of equations, they are called *mathematical models*.

The so-called *laws of Nature* are not, for the modern scientist, laws of the real world but of the space of idealizations. Basic elements of a theory are, therefore, that space and the transfer function mapping each object or prototype to its idealization or model. Two prototypes are said to be equal, in the light of a given theory, if in the space of idealizations they have the same model.

It is the aim of this article to show how scientific progress, implemented by advances in the technique of modelling, as a rule, began by introducing into Engineering a certain impoverishment which makes the next step a necessary one.

Accordingly to Ortega y Gasset, in order to prevent that from happening, "on entering in the profession, Science has to be disarticulated as Science in order to be organized, according to another centre or principle, as a professional technique".

Only in this way it is possible to prevent Engineering, fecundated by Science, from losing its specific character, from ceasing to be Engineering.

3. MODELLING OF THE GEOMETRICAL FORMS

One of the first scientific problems that engineers had to face was that of describing the structures they intended to construct, from a geometrical point of view.

Constraints connected with the materials used and the construction processes available forced them to base that description on a relatively limited set of lines and surfaces of simple and double curvature. The convenience of using *voussoirs* that were all equal in fact led to circular directrices. That was a reason why parabolic directrices, for instance, were excluded.

It was in the corresponding space of idealizations that the medieval builders resolved the geometrical problems associated with the covering of large spans. The solutions they found were the basis for several architectural styles.

Such was the case when the solution arrived for the problem of covering a square plan with a spherical cupola, making exclusive use of spherical surfaces, which formed the basis for the Byzantine architecture.

The introduction of directrices consisting of two circular arches, i.e. ogival directrices, extended the range of possibilities and later enabled Western architects to resolve the problem of covering a plan which was no longer square but rectangular, with two pointed-arch vaults intersecting at right angles.

In both cases it is clear that the adopted space of idealizations conditioned the evolution of the engineering of stone structures. It was necessary to wait for new materials to appear before the shape grammars of builders could be enriched.

4. MODELLING OF THE MECHANICAL BEHAVIOUR OF STRUCTURES

The builders of former times resorted to empirical rules for ensuring the stability of their constructions. Those rules were used together with others whose significance was merely aesthetic.

An example from Alberti's rules for stone bridges, quoted in his XV century book "De re aedificatoria" (Alberti, 1485), may be given: "The width of the piers should be one quarter of the height of the bridge; the clear span of the arch should not be more than six times, and not less than four times the width of the piers; the thickness of the *voussoirs* should be not less than 1/10 of the span".

Builders followed those principles of the art as nowadays engineers comply with regulations and codes of practice. The great difference lies in the fact that regulations constitute the final product of a whole series of so called pre-normative research, whereas the ancient rules lacked a scientific basis. Such was the situation that led to the development of mechanical models of the behaviour of structures, i.e., to the modern engineering of structures.

That turning point was not accepted by the profession, either immediately or without argument.

In England, for instance, unlike what occurred in France and Germany, during a good part of the 19th Century there persisted the situation - by then anachronistic - of a structural engineering practice that was almost completely ignorant of mechanics.

In the United States, a country where scientists nowadays are so greatly stimulated by the opportunities offered to them for applying the results of their studies, the situation was quite different from what it is now, nearly up to the Second World War. Timoshenko's memoirs clearly show this (Timoshenko, 1968).

It is only fair to recognise that the application of mechanics to engineering did not bring only advantages: engineers began, in fact, to construct almost exclusively those structures which they knew how to analyse and to base their calculations on principles so simplistic that, according to them, a simple factory chimney stack, of the kind made of bricks, should not be entitled to be standing.

5. MODELLING OF ANALYSIS

The appearance of computers in the middle of the 20th Century caused in structural engineering what can be called the second revolution (the first coinciding with the use of mechanics).

The new possibilities enabling engineers to make more extensive use of existing scientific tools, allowed them to access to fields that had previously been considered to lie outside their spheres of interest.

It has been seen that, after the first revolution, engineers started to construct exclusively those structures that they knew how to analyse. That drawback ceased to exist, however, since the differential equations of the various structural models could be the object of numerical processing, that is to say, since all structures could be analysed.

On the other hand, computers brought with them another possibility: that of computing automation, which making it necessary the modelling and programming of analysis, in turn raised new problems.

In an article published in 1964, Charles Miller, then Head of the Civil Engineering Department of the MIT, made an analysis - now historic - of the perverse effects that the use of computers was then having on civil engineering.

According to him, a well established pattern of computer usage had been built, during the preceding decade, on the concepts of the program and the program library, an outgrowth of traditional methods of programming which called for skilled specialists (programmers), great investment in time and money, and well defined program situations. Based on these concepts, a whole generation of users, computer staff people and organisation managers had developed new mental attitudes and operational mechanisms. On the other hand, hundreds of computer centres had been established and were being operated on such principles in civil engineering organisations.

During the same period, numerous doctrines regarding problem requirements for computer usage were established. The requirements repeated most often had indicated that, to justify the use of a computer on a problem, the problem should be highly repetitive, occurring often enough in the same form to permit using the program many times, and complex enough to involve computational effort that would be prohibitive by manual methods. Computer libraries were full of the first type of problem; journals and magazines were increasingly full of papers based on the second type.

Now, the development of a classical program involved such well-known steps as problem definition, mathematical and logical analysis, flow charting, programming, coding, debugging, testing and documenting. These steps often required months and workload and impressive financial investments. Because the development of a useful program was such a major task it rather completely controlled how and to what extent computers could be effectively used in civil engineering. As computer libraries developed, the question of computer usage was decided on the basis of whether or not a program was available in the library which matched the problem situation.

Miller's conclusions was that the most valid criticism to a classical computer usage was the attempt to force engineering to conform to the input requirements of an available library program. According to him, there was evidence that expensive library programs might become real obstacles to technical and professional progress. Namely, the routine filling out of standard forms for rigid computer programs was completely distasteful and uninteresting to the engineer. As a result, many engineers had developed the mental attitude that computers were only applicable to routine, non-creative problems. Despite the tremendous publicity given to computers, and despite the impression that all was well, it should be acknowledge that the actual result had fallen considerably short of expectations.

Charles Miler went on, then, into details about the role and functions that should be attributed to engineers. The engineer being essentially a decider, one of the most important characteristics of the decision-maker process in engineering is the fact of its being based on imprecise and incomplete information, since the cost of complete information, even if available, must be prohibitive. Moreover, variables and non-quantifiable factors assume great importance in engineering. Decisions are to be taken, however, under the pressure of time.

Apart from this, what is asked to the engineer is an optimal solution constrained from points of view that are far from being merely technical.

Regarding an engineering organisation as an information system receiving raw information which has to be reduced, processed, stored and combined with other information, Miller pointed out that technicians were given tasks of this type which could and should be shared with computers. One might ask for what reasons was it difficult to hand over such tasks to computers. Some plausible reasons might be that it was far easier to give instructions to a human being than to a computer, or that, since the time that counts is not only the computation time but also the time of access to the computer, a human being could often, in practice, respond far more quickly than a machine.

Charles Miller concluded that computers could only come up to the expectations which they had created in the engineering community if communications between man and machine were at least as efficient as communication between engineers[1] and access to computers were considerably facilitated.

The latter condition is nowadays largely met, while many contributions towards the former are also resolved. In other words, it became possible to automate a series of processes that formerly relied strictly on human intervention. Hence, the need to study those processes, which naturally led to considering their appropriate modelling.

6. MODELLING OF DESIGN PROCESSES

The modelling of knowledge-based processes – such as design – and, in particular, its modelling within an engineering environment, may be said to correspond to a recent endeavour, largely resulting from the extension of early attempts to enhance computer-aided engineering tasks, confusingly gathered under the CAD acronym.

These would vary from sophisticated draughting programs, through solid modelling systems, to powerful engineering analysis packages.

Draughting programmes are of undeniable value as useful tools in representing the end result of a design, but can hardly sustain themselves as comprehensive design assistants as much as solid modelling systems could not.

Engineering analysis packages for structural analysis or structural optimization, covered no more than specific design sub problems that may could be clearly described by well established governing laws through the use of mathematical models.

These systems could partially help the engineer in particular stages of the design process, but, as Charles Miller rightly stressed, could not be used to address the full process of design. Basically, they are calculating programs through which a set of numbers describing the

---

[1]   This concern recalls an interesting resemblance with the Touring test, put forward to establish a means to evaluate "computational intelligence" (Turing, 1963).

problem - the input - is mapped to another set of numbers describing the results - the output - through the pre-established use of a set of procedures known as an algorithm (Fenves, 1987).

The inability of such systems to address the whole design process, and in particular the lack of overall support for decision making, may be explained, at least partly, by the following:

- from the software engineering stand-point, the inadequacy of the then available programming paradigms to represent and manipulate information of a non-numerical nature - such as approximations, rules of thumb and more general heuristics, or lines of reasoning, which constitute essential but ill-structured knowledge components of most design problems;

- from the commercial point of view, the short term success strategy pursued by the computer industry has conflicted with the long term nature associated with the development of more comprehensive systems.

However, this conventional concepts of CAD and computer-aided engineering (CAE) systems have rapidly been changing because engineers' needs and expectations evolved towards more sophisticated requirements, but also due to the opportunities created by the emergence of new technologies.

Three sets of reasons may be associated with such new trends in computer-aided engineering:

- the amount of information and knowledge of different sources - scientific, legislative, technical, etc. - required to be processed and dealt with in engineering environments has reached a level too extensive to be handled by engineers alone;

- the urge to use comprehensive design-aiding systems, suitable for assisting engineers intelligently during the whole decision-making process, from specification of goals to delivery of objects;

- the emergence of new contributions from computer science, engineering design research and cognitive science, allowing a higher functionality to be eventually achieved by CAD systems; particularly relevant is the advent of *artificial intelligence* (AI) , namely through *expert* systems technology, which has been considered a primary candidate to participate in the building of a new generation of CAE systems (ten Hagen and Tomiyama, 1987).

A similar relevance may be attached to the rise of design theories and models based on work from cognitive scientists, the development of which was driven by the cognitive needs of the designer, as opposed to former approaches (see Miller, 1963) favouring the satisfaction of ergonomic requirements, such as the use of graphical/iconic interfaces of interactive input devices.

New CAE systems should be able not only to execute tiresome and routine tasks but also to perform mundane problem solving activities while requiring no user intervention.

This should enable engineers to concentrate on the crucial and typically human-dependent activities in engineering, such as synthesis; for these reasons such systems have been classified as intelligent.

The level of non-human synthesis which these systems can perform, and the mundaneness of the sub problems they can solve, are related with, and provide a measure of the degree of intelligence they possess (Bento, 1992).

A new generation of such systems has now been gradually specified and developed at a research and commercial level; they assume disparate forms and a variety of names, for instance, *intelligent CAD systems - ICAD, knowledge-based CAD systems* and *CAD experts systems*, but have been also referred to as *design automation systems*.

In this context, design automation, meaning the substantial replacement, during design, of human action by that of a computer, without removing the human interference or, at least, the human control in the key stages of the process, has emerged as the common concept between these new approaches in CAD.

7. CLOSURE

Most current research on more intelligent and comprehensive CAD systems would appear to be concerned with furthering automation of the design process by incorporating more of the traditional user's role into the system. However, the design automation goal requires a deeper understanding of the whole process of design (Feijó, 1988) and sufficient theory has not yet been developed to support fully functional computable models of the design process.

The formulation of such computable models should not be faced by the engineering community as a goal in itself, but rather as a fundamental requirement to achieve a more solid foundation on which the emerging intelligent systems will have to rely.

It is well known tha the modelling of engineering design processes, as well as of many other knowledge-based processes, is a very complex task; if not for other reasons, this is so because the modelling of human reasoning is manifold: it includes activities based on *induction, deduction, abduction, intuition, experience* and *creativity* (among possibly many others).

These activities are difficult to formalise and, although some of the currently available computational tools provide the means for its computational modelling (for example: logic programming adequately implements the process of deduction; knowledge-based systems (KBS) are successfully used for implementing some features of experience, etc.), the modelling of knowledge-based processes is still on its infancy, and facing serious limitations.

The heart of the matter is that, algorithms being the vehicle for representing human knowledge, the range of activities that can be performed by a computer is confined to those tasks for which humans can find algorithms.

The following argument has been used to establish the distinction between information processing done by computers and human information processing: "humans are capable of developing their behaviour through <u>learning</u>, while computers have to wait for some human to feed them the algorithms required to accomplished the desired task" (Alexander and Morton, 1990).

The scientific issue that seems to be in line for further progressing in engineering seems to be, therefore, that of *modelling the learning process*, for which the symbolic processing paradigm may not be the most adequate one. *Connectionism* will, therefore, have a role to play in CAE (as illustrated later in the book).

Moreover, less rigid and less formal models of the design process are also in line for further research. Indeed, concepts such as *virtual design environments* and *active support* to design tasks will dictate much of the next steps in this area, as seen throughout the book.

In fact, most existing approaches to ICAD lack direct support to cognitive needs that are associated to the design task environment. Also, traditional KBS approaches to ICAD systems, especially those strictly based on logic and problem solving techniques, require a modelling of the design process associated with precise, rigid, discrete and non-parallel thought processes, whose implementation require a great deal of anticipation, while, in fact, design thought processes are often vague, fluid, ambiguous and amorphous.

In addition to many current approaches to ICAD for structural engineering that can be described as problem solving systems for design, eventually guided by the user, newer and eventually more effective ones should enable higher interactivity, directly coping with designer's cognitive needs and providing forms of active support as introduced by Smith (1996).

Such less formal models and more (pro-)active systems may be pursued by endorsing, novel paradigms such as that of *agents*.

REFERENCES

ALBERTI, L. B, **1964**: *De Re Edificatoria* , Florence, 1485, quoted in Straub, H.- *A History of Civil Engineering* , The MIT Press, USA.

ALEKSANDER, I.; MORTON, H., **1990**: *Neural Computing*, Chapman & Hall, London, UK.

ARANTES E OLIVEIRA, E. R.; PEDRO , J. O, **1986**: "The rise and decline of structural analysis as a research topic in civil engineering", J.S.S.T. Conference on Recent Advances in the Simulation of Complex Systems, Tokyo, Japan.

BENTO, J. P., **1992**: *Intelligent CAD in Structural Steel: a Cognitive Approach*, PhD Thesis, Expert Systems Laboratory, Department of Civil Engineering, Imperial College of Science, Technology and Medicine, London, UK.

FEIJÓ, B., **1988**: *Fundamental steps towards an intelligent CAD system in structural steel*, PhD Thesis, Expert Systems Laboratory, Department of Civil Engineering, Imperial College of Science, Technology and Medicine, London, UK.

FENVES, S. J., **1987**: "Expert systems in civil engineering – State of the art", in *Fourth International Symposium on Robotics and Artificial Intelligence in Building Construction*, Haifa, Israel.

MILLER, C. L., **1963**: "Man-machine communication in civil engineering", Proc. ASCE, J. Str. Div., p. 3593, USA.

SMITH, I., **1996**: "Interactive Design - time to bite the bullet", *Information Processing in Civil and Structural Engineering Design*, B. Kumar (Ed.), 23-30, Civil-Comp Press, Scotland.

TEN HAGEN, P.; TOMYIAMA, T. **1987**: "Preface", in ten Hagen, P. J. W.; Tomiyama (eds.) *Intelligent CAD Systems I: theoretical and methodological aspects*, Springer-Verlag, Berlin, Heidelberg, New York.

TIMOSHENKO, S.P., **1968**: *As I remember: the autobiography of Stephen P.Timoshenko*, van Nostrand, Princeton, NJ

TURING, A., **1963**: "Computing Machinery and Intelligence", in E.A. Feigenbaum and J.Feldman, eds. *Computers and Thought*, McGraw-Hill, New York.

AN INTRODUCTION TO ARTIFICIAL INTELLIGENCE AND TO THE DEVELOPMENT OF KNOWLEDGE-BASED SYSTEMS

C. Tasso

University of Udine, Udine, Italy

1. Artificial Intelligence

Artificial Intelligence is the field of computer science aimed at developing hardware and/or software systems[1] (more generally, computational models) capable of performing functions which have been traditionally considered unique and exclusive of human cognition [Bundy 80], [Nilsson 80], [Barr et al. 81-89], [Pearl 84], [Charniak and McDermott 85], [Boden 87], [Ford 87], [Shapiro 87], [Banerji 90], [Schalkoff 90], [Rich and Knight 91], [Winston 92]. Among those capabilities, we can mention natural language and speech processing, expert reasoning and problem solving, such as diagnosis and design, vision, and learning.

The 'artificial intelligence approach' to the development of computer programs is inherently different from the so called traditional approach. In the traditional approach, when a problem has been identified, the programmer has to design a suitable algorithm for solving the problem, to write the algorithm by means of a suitable programming language, and to finally load the program into the computer for its execution: the computer acts solely as an executor of the program received in input. The artificial intelligence approach is organized differently: the goal is to provide the computer with problem solving capabilities. As such, the computer will be able to receive the description of a problem in a given domain, and autonomously find the solution method for the problem at hand. However, achieving this ambitious goal requires to 'augment' the computer with specific domain *knowledge*, which enables the computer to analyze problems in that domain and find specific reasoning paths towards their solution, with more general, powerful, and flexible results than what is normally obtained with the traditional approach.

Knowledge plays a central role in artificial intelligence: it must be identified, acquired, appropriately represented, and processed in order to show intelligent problem solving behavior. For this reason, *Knowledge Acquisition* and *Knowledge*

---

[1] In this paper we will restrict our attention only to software systems.

Representation are two central areas of investigation in artificial intelligence. The former deals with the methods and techniques that are to be exploited by a *knowledge engineer* (i.e., the professional role in charge of developing an artificial intelligence system) in order to identify appropriate knowledge sources, to elicit from them relevant domain knowledge, and to build abstract models of the knowledge and reasoning methods exploited in the specific domain at hand. Knowledge acquisition has received great attention form the research community, and a huge amount of scientific and technical publications have been produced. Knowledge acquisition methods are out of the scope of this short introduction, and the interested reader can refer to [Guida & Tasso 94] for an extensive bibliography and the presentation of a specific methodology for knowledge acquisition within the development of knowledge-based systems.

Knowledge representation is another fundamental corner stone of artificial intelligence: it is aimed at developing suitable formal languages (called *knowledge representation languages*) adequate for coding all the knowledge relevant for a given application. A knowledge representation language is characterized by a syntax, specifying how the symbols of the language can be correctly used and combined together, and a semantics, which describes their meaning in terms of domain knowledge. Moreover, a knowledge representation language is characterized by a set of *reasoning algorithms*, i.e. algorithmic procedures which specify how the specific knowledge representation language can be manipulated in order to automatically perform problem solving and reasoning activities which resemble human problem solving behavior. A knowledge representation language should feature *expressiveness*, i.e. the capability to represent all the relevant knowledge of a domain with the adequate level of precision; *efficiency*, i.e. the capability to minimize the computational resources needed to represent knowledge and to reason with it; and, last but not least, *cognitive adequacy*, i.e. the capability to represent knowledge and to perform reasoning in a human-like fashion, thus enabling a human observer to naturally discover in the artificial performance the nuances and traits of human ways of dealing with knowledge. The most common knowledge representation languages are: logic formalisms [Genesereth and Nilsson 87], semantic networks [Findler 79], frames [Minsky 75], and production rules [Davis and King 77]. A detailed illustration of knowledge representation and reasoning mechanisms is out of the scope of this paper. The interested reader may refer to [Laubsch 84], [Ringland and Duce 84], [Sowa 84], [Brachman and Levesque 85], [Charniak et al. 87], [Bench-Capon 90], [Schalkoff 90], [Davis et al 93].

2. Knowledge-Based Systems: Basic Definitions

This section and the following ones are synthesized from the volume [Guida and Tasso 94].

A *knowledge-based system* (*KBS*) is a software system capable of supporting the explicit representation of knowledge in some specific competence domain and of exploiting it through appropriate reasoning mechanisms in order to provide high-level problem-solving performance [Hayes-Roth et al. 83], [Addis 85], [Waterman 86], [Rolston 88], [Guida and Tasso 89], [Chorafas 90], [Harmon and Sawyer 90], [Jackson 90], [Prerau 90], [Edwards 91], [Guida and Tasso 94]. Therefore, a KBS is a specific, dedicated, computer-based problem-solver, able to face complex problems, which, if solved by humans, would require advanced reasoning capabilities, such as deduction, abduction, hypothetical reasoning, model-based reasoning, analogical reasoning, learning, etc.

From an abstract point of view, a KBS is composed of two parts:
- a central part, called *kernel.*, which implements the basic problem-solving capabilities of the KBS;
- a peripheral part which is aimed at providing additional functions necessary for a practical and effective use of the KBS and comprises a collection of *special-purpose modules*.

The kernel is constituted by three parts: the knowledge base, the reasoning mechanism, and the working memory. The *knowledge base* stores available knowledge concerning the problem domain at hand, represented in an appropriate explicit form through a knowledge representation language and ready to be used by the reasoning mechanism. It may contain knowledge about domain facts and objects, their structure and properties, the relations existing among them, the structure of typical problems, and the problem-solving strategies. The knowledge base is a highly structured, long-term memory, which can store knowledge permanently during the whole life-time of the KBS. The *reasoning mechanism* is constituted by a complex set of programs capable of performing high-level reasoning processes in order to solve problems in the considered domain by exploiting the knowledge stored in the knowledge base. The *working memory* is used to store all information relevant to a problem solving session of the KBS, that is: a description of the problem at hand, the intermediate solution steps, and eventually the solution found. The working memory is a short-term memory which is updated each time a new problem is considered. Its content lasts as long as the problem-solving session does.

At a high level of abstraction the operation of a KBS kernel may be schematically described through the high-level procedure illustrated in Figure 1, which implements the so-called *basic recognize-act cycle*.

From another perspective, we may consider a KBS kernel as having two parts:
1. a container, called the *empty system kernel*, an algorithmic part constituted by the complex set of programs defining the structure and organization of the knowledge base and of the working memory of a KBS and implementing its reasoning mechanism;

```
begin
  INSERT in working memory a representation of the problem to solve
repeat
     SEARCH in the knowledge base for knowledge potentially relevant to the solution
             of the current problem
     if any useful knowledge is not found then EXIT LOOP with failure
     SELECT from the set of retrieved knowledge the knowledge to be used
             in the current cycle
     APPLY the selected knowledge and possibly transform the current content
             of the working memory
until problem solved
end.
```

Figure 1. Basic recognize-act cycle of a KBS [Guida and Tasso 94].

2. a content, made up by the actual, specific knowledge stored in the knowledge base.

The first part is general and mainly application independent, while the second part strictly depends on the specific application considered. The empty system can be viewed as an abstract, general problem-solver which can be used to solve problems in a specific application domain through the insertion of appropriate knowledge in the initially empty knowledge base.

The special-purpose modules of a KBS include:
- A *software interface* which connects the KBS to external software systems.
- An *external interface* which connects the KBS to the external environment in which it operates, such as sensors, data acquisition systems, actuators, etc.
- A *user interface* which is aimed at making the interaction between the user and the KBS friendly and effective. It may include advanced man-machine interaction facilities, multimedia and dialogue systems, user modeling [Jameson et al. 97] and smart help systems.
- An *explanation system* which is directly connected to the knowledge-based components of the KBS and explains and justifies the behavior of the KBS to the user, by showing the knowledge utilized, the problem-solving strategies used, and illustrating the main reasoning steps.

More and more, KBSs are not developed in isolation, but, on the other hand, they are variously integrated or embedded with traditional software systems. In engineering, these other systems may concern CAD/CAM packages, libraries of specific processing routines, production management and, more generally, information systems, and so on.

For the above reason, traditional practice in software engineering provides a valuable, general, and effective corpus of knowledge to be applied for software development projects including a KBS component. However, KBSs feature unique characteristics, which motivate more specific and focused development techniques and methodologies. In the next section, a life cycle model specifically conceived for KBS development will be shrtly presented.

3. KLIC: A Life Cycle Model for the KBS Development

KBS development is a complex issue, and there are often many alternative ways to organize and perform the various tasks involved in the design, production, maintenance, and extension of a knowledge-based product. The term *KBS life cycle* refers to how the various tasks involved in the development of a KBS are defined and organized.More precisely, the life cycle is a reference schema which specifies [Boehm 88] [Hilal and Soltan 93]:

- what to do, i.e. which specific tasks should be performed, and
- when to do it, i.e. in which contexts, under what conditions, and in which temporal and logical order.

In the following the *KLIC* (*Kbs LIfe Cycle*) [Guida and Tasso 94] concept of the life cycle for KBS development is surveyed. It is structured into six *phases*, corresponding to the major stages of the analysis, design, production, and maintenance process. Each phase is decomposed into a reasonably limited number of *tasks*, grouped into *steps*.

For a detailed description of KLIC the interested reader can refer to [Guida and Tasso 94].

3.1 Phase 0: Opportunity analysis

Before starting any specific KBS project in a given organization, it is generally appropriate to develop a broad-spectrum investigation to locate, evaluate, and rank opportunities for KBS applications in the organization. This is the subject of *opportunity analysis*. More specifically, this phase identifies within a given organization (company, institution, government department, etc.) the application areas which could benefit from the development of KBS projects, and ranks them according to their strategic value, tactic importance, expected benefits, technical complexity, suitability and readiness for KBS application, involved risk, logical and temporal precedence, etc. This phase does not strictly belong to the life cycle of a specific KBS, but it constitutes a sort of background study which precedes the development of several specific KBS projects.

The main product of opportunity analysis is the *master plan*, a coarse-grained plan used as a long-term reference to guide an organization in the most appropriate application and exploitation of knowledge-based technology. When a specific KBS project is then initiated, the master plan can suggest the most appropriate application area to focus on, thus providing a useful input to phase 1.

Opportunity analysis is not a mandatory phase of the life cycle, even if it is highly recommended, except for organizations of small dimensions and complexity.

Opportunity analysis comprises 15 tasks organized into 4 steps, as illustrated in Table 1.

● START-UP		
	0.1	verification of prerequisites
IF		prerequisites are not satisfied
THEN	0.2	writing of management report
		stop phase
	0.3	planning of opportunity analysis
● ANALYSIS OF THE EXISTING ORGANIZATION		
	0.4	analysis of objectives
	0.5	process and structure analysis
PARALLEL		
	0.6	analysis of the level of automation
	0.7	identification of areas and domains
END-PARALLEL		
● ANALYSIS OF OPPORTUNITIES		
	0.8	characterization of domains
	0.9	identification of knowledge problems and definition of potential KBS applications
	0.10	definition and characterization of opportunities
● SYNTHESIS AND RELEASE		
	0.11	construction of the master plan
	0.12	writing of draft opportunity analysis report
	0.13	presentation and acceptance of results
IF		revision is needed
THEN	0.14	revision of opportunity analysis
	0.15	writing of final opportunity analysis report

Table 1. Task structure of the KLIC 'opportunity analysis' phase
[Guida and Tasso 94].

3.2 Phase 1: Plausibility study

The *plausibility study* encompasses the following main goals:
- analyzing a given application domain - possibly suggested by the master plan, and identifying a specific problem to face;
- analyzing the requirements and defining the overall project goals;
- identifying the main functional, operational, and technical specifications of the KBS, and the acceptance criteria;
- developing a draft technical design, a draft organizational design, and a draft project plan;
- assessing the global plausibility of the KBS application.

The concept of plausibility includes five aspects, namely: technical feasibility, organizational impact, economic suitability, practical realizability, and opportunities and risks. A positive evaluation of plausibility requires that all such aspects receive a positive independent evaluation.

The product of the plausibility study is the *plausibility study report*. It is a technical document for the management which illustrates the activities done and the results obtained, it suggests choices and decisions about the KBS project, and it proposes a draft system design and a draft project plan.

Plausibility study is a mandatory phase of the life cycle. It comprises 29 tasks organized into 5 steps, as illustrated in Table 2.

- **START-UP**

	1.1	verification of prerequisites
IF		prerequisites are not satisfied
THEN	1.2	writing of management report
		stop phase
	1.3	planning of plausibility study

- **INITIAL ANALYSIS**

	1.4	process and structure analysis
PARALLEL		
	1.5	analysis of the level of automation
	1.6	verification of domain
END-PARALLEL		
	1.7	characterization of domain
	1.8	identification of knowledge problems and verification of potential KBS application
	1.9	characterization of potential KBS application
IF		initial analysis does not conclude with a positive evaluation
THEN	1.10	writing of management report
		stop phase

- **BASIC DEFINITIONS**

	1.11	analysis of requirements
	1.12	definition of project goals
PARALLEL		
	1.13	definition of functional specifications
	1.14	definition of operational specifications
	1.15	definition of technical specifications
END-PARALLEL		
	1.16	definition of acceptance criteria

Table 2-a. Task structure of the KLIC 'plausibility study' phase: the first three steps [Guida and Tasso 94].

3.3 Phase 2: Construction of the demonstrator

The main goal of the *construction of the demonstrator* is to develop and demonstrate a first, limited version of the KBS in order to meet one or more of the following issues:

- obtaining a concrete insight in the complexity of the problem considered, and validating, refining, and, if necessary, revising technical decisions outlined in the plausibility report;
- validating, refining, and, if necessary, revising the draft project plan developed in phase 1;

- **ANALYTICAL ASSESSMENT OF PLAUSIBILITY**

PARALLEL		
	BEGIN	
	1.17	assessment of technical feasibility
	1.18	development of draft technical design
	END	
	BEGIN	
	1.19	assessment of organizational impact
	1.20	development of draft organizational design
	END	
END-PARALLEL		
	1.21	development of draft project plan
PARALLEL		
	1.22	assessment of economic suitability
	1.23	assessment of practical realizability
	1.24	assessment of opportunities and risks
END-PARALLEL		

- **SYNTHESIS AND RELEASE**

	1.25	global evaluation of plausibility
	1.26	writing of draft plausibility study report
	1.27	presentation and acceptance of results
IF		revision is needed
THEN	1.28	revision of plausibility study
	1.29	writing of final plausibility study report

Table 2-b. Task structure of the KLIC 'plausibility study' phase: the final two steps [Guida and Tasso 94].

- collecting useful feedback from the users, and refining the identification of requirements and the definition of KBS specifications stated in phase 1;
- gaining involvement and commitment from the management or from the potential client;
- securing the interest and cooperation of experts and users, that will be crucial in later phases of the project.

The products of this phase are:
- a running KBS, called *demonstrator*, which anticipates the KBS performance on a limited part of the considered problem;
- the *demonstrator report*, which contains a synthesis of the activities carried out and a detailed illustration of the results achieved.

Note that, according to the goals actually considered, several types of demonstrators may be possible, including promotional, commercial, involvement, exploratory, experimentation, organizational, and planning.

Construction of the demonstrator, although common in several KBS projects, is not a mandatory phase of the life cycle, and it is one of the objectives of a plausibility study to suggest whether a demonstrator needs to be developed or not.

Construction of the demonstrator comprises 18 tasks organized into 5 steps, as illustrated in Table 3.

- **START-UP**

	2.1	verification of prerequisites
IF		prerequisites are not satisfied
THEN	2.2	writing of management report
		stop phase
	2.3	planning of construction of the demonstrator

- **BASIC CHOICES AND DEFINITIONS**

2.4	identification of demonstrator goals
2.5	identification of sub-problems and sample cases
2.6	definition of demonstrator specifications

- **ANALYSIS AND DESIGN**

2.7	conceptual modeling
2.8	selection and acquisition of the basic development environment
2.9	technical design

- **DEVELOPMENT**

2.10	implementation of the empty system
2.11	implementation of the development support system

LOOP		
	2.12	knowledge acquisition planning
	2.13	knowledge elicitation and protocol analysis
	2.14	knowledge coding
	2.15	knowledge integration and verification
UNTIL		demonstrator specifications are fully met

- **DEMONSTRATION, EVALUATION, AND SYNTHESIS**

2.16	demonstration
2.17	evaluation
2.18	writing of demonstrator report

Table 3. Task structure of the KLIC 'construction of the demonstrator' phase
[Guida and Tasso 94].

3.4 Phase 3: Development of the prototype

The *development of the prototype* is the main endeavor of a KBS project. Its main objective is to find the most suitable technical solutions for the application considered, and to implement these in a running system.

The products of this phase are:

- a full KBS, called *prototype*, which can adequately meet all functional specifications stated;
- an integrated set of software tools, called *development support system*, which supports the construction of the knowledge base of the prototype;

- the *prototype report*, which contains a synthesis of the activities carried out and a detailed illustration of the results achieved.

The prototype, although satisfying the functional specifications stated, is not the final output of the production process, since:
- it is not yet installed in the real operational environment, but it is running only in the development environment (if necessary, connections with the external world are simulated);
- it has only been tested with sample data prepared by the system designer with the support of experts and users;
- it is still embedded in the development environment and it is neither engineered nor optimized.

Let us point out that the prototype is generally a completely different system from the demonstrator. Only very seldom can the prototype be obtained from the demonstrator through appropriate extension and refinement. This is motivated by the fact that the objectives, the design principles, and the development tools used for the two systems are definitely different.

Development of the prototype comprises 22 tasks organized into 6 steps, as illustrated in Table 4.

- **START-UP**

	3.1	verification of prerequisites
IF		prerequisites are not satisfied
THEN	3.2	writing of management report
		stop phase
	3.3	planning of development of the prototype

- **CONCEPTUAL DESIGN**

	3.4	knowledge analysis and modeling
	3.5	design of the conceptual model
	3.6	definition of prototype specifications

- **TECHNICAL DESIGN**

	3.7	design of the logical model
IF		shortcomings in knowledge analysis and modeling are identified
THEN		**GO TO** 3.4
	3.8	definition of the specifications of the empty system
	3.9	definition of the specifications of the development support system
	3.10	selection and acquisition of the basic development environment
	3.11	detailed design of the empty system
	3.12	detailed design of the development support system

- **CONSTRUCTION OF THE EMPTY SYSTEM AND OF THE DEVELOPMENT SUPPORT SYSTEM**

	3.13	implementation of the empty system
	3.14	implementation of the development support system

Table 4-a. Task structure of the KLIC 'development of the prototype' phase: the first four steps [Guida and Tasso 94].

- **DEVELOPMENT OF THE KNOWLEDGE BASE**

	3.15	selection of knowledge sources
LOOP		
	3.16	knowledge acquisition planning
	3.17	knowledge elicitation and protocol analysis
	3.18	knowledge coding
	3.19	knowledge integration and verification
IF		revision of empty system is necessary
THEN		**GO TO** 3.7
IF		revision of prototype specifications is necessary
THEN		**GO TO** 3.6
UNTIL		prototype specifications are fully met

- **TESTING AND EVALUATION**

	3.20	prototype testing and refinement
	3.21	prototype evaluation
	3.22	writing of prototype report

Table 4-b. Task structure of the KLIC 'development of the prototype' phase: the final two steps [Guida and Tasso 94].

3.5 Phase 4: Implementation, installation, and delivery of the target system

The goal of *implementation, installation, and delivery of the target system* is to develop a complete KBS. It must have the same behavior of the prototype, but in addition it must be:

- installed in the real operational environment;
- field tested with real data;
- engineered and optimized;
- eventually delivered to the end-users for routine operation.

The products of this phase are:

- the *target system*, that is, the final output of the whole KBS production process;
- the *maintenance support system*, that is the specific system devoted to support effective and efficient maintenance;
- the complete set of manuals, including user, maintenance, and technical manuals, necessary for correct and effective system operation;
- the *target system report*, which contains a synthesis of the activities carried out and a detailed illustration of the results achieved.

The implementation of the target system may require very different approaches, depending on the requirements and constraints imposed by the operational environment, ranging from the automatic generation of the delivery version through specific support tools, to the incremental refinement and engineering of the prototype, to complete re-implementation - in the most unfortunate cases. Of course, each approach entails different production plans, costs, time, and technical features of the obtained target system.

Implementation, installation, and delivery of the target system comprises 20 tasks organized into 5 steps, as illustrated in Table 5.

- **START-UP**

<table>
<tr><td></td><td>4.1</td><td>verification of prerequisites</td></tr>
<tr><td>**IF**</td><td></td><td>prerequisites are not satisfied</td></tr>
<tr><td>**THEN**</td><td>4.2</td><td>writing of management report
stop phase</td></tr>
<tr><td></td><td>4.3</td><td>planning of implementation, installation, and delivery of the target system</td></tr>
</table>

- **PREPARATION**

<table>
<tr><td></td><td>4.4</td><td>analysis of target environment</td></tr>
<tr><td></td><td>4.5</td><td>definition of specifications of the target system and of the maintenance support system</td></tr>
<tr><td colspan="3">**PARALLEL**</td></tr>
<tr><td colspan="3">**BEGIN**</td></tr>
<tr><td></td><td>4.6</td><td>definition of the approach for target system implementation</td></tr>
<tr><td></td><td>4.7</td><td>design of the target system and of the maintenance support system</td></tr>
<tr><td colspan="3">**END**</td></tr>
<tr><td></td><td>4.8</td><td>organizational design</td></tr>
<tr><td colspan="3">**END-PARALLEL**</td></tr>
</table>

- **PRODUCTION AND INSTALLATION**

<table>
<tr><td colspan="3">**PARALLEL**</td></tr>
<tr><td></td><td>4.9</td><td>production of the target system and of the maintenance support system</td></tr>
<tr><td></td><td>4.10</td><td>organizational intervention</td></tr>
<tr><td colspan="3">**END-PARALLEL**</td></tr>
<tr><td></td><td>4.11</td><td>installation</td></tr>
<tr><td></td><td>4.12</td><td>field testing and refinement</td></tr>
</table>

- **FIRST RELEASE AND EXPERIMENTAL USE**

<table>
<tr><td>4.13</td><td>writing of draft manuals</td></tr>
<tr><td>4.14</td><td>training of users</td></tr>
<tr><td>4.15</td><td>first release and experimental use</td></tr>
</table>

- **FINAL RELEASE**

<table>
<tr><td>4.16</td><td>refinement of the target system and of the maintenance support system</td></tr>
<tr><td>4.17</td><td>writing of final manuals</td></tr>
<tr><td>4.18</td><td>certification and acceptance</td></tr>
<tr><td>4.19</td><td>final release</td></tr>
<tr><td>4.20</td><td>writing of target system report</td></tr>
</table>

Table 5. Task structure of the KLIC 'implementation, installation, and delivery of the target system' phase [Guida and Tasso 94].

3.6 Phase 5: Maintenance and extension

Maintenance and extension starts after the delivery of the target system to the users for operational use and lasts for the entire life of the KBS. This phase is of primary importance and its goal is to ensure a long and effective operational life of the KBS and to exploit all potential benefits of the project.

Maintenance only involves modifications to the knowledge base of the KBS, while extension requires changes to be made even to the fundamental structures of the empty system.

The products of this phase are:

- new versions of the knowledge base of the target system and updated manuals, concerning maintenance;
- new versions of the target system and of the maintenance support system, and updated manuals, concerning extension;
- the *KBS history*, an evolving document which includes the collection of *operation reports* collected form the users, containing their feedback, remarks, and requests, and a record of all activities carried out during maintenance and extension.

Implementation, installation, and delivery of the target system comprises 18 tasks organized into 7 steps, as illustrated in Table 6.

- **PREPARATION**

	5.1	definition of the strategy for maintenance and extension
	5.2	training of the maintenance team

LOOP
- **OBSERVATION**

LOOP		
	5.3	collection of operation reports
UNTIL		a meaningful number of operation reports has been collected

- **INTERVENTION SET-UP**

	5.4	analysis of operation reports and verification of prerequisites
IF	prerequisites are not satisfied	
THEN	BEGIN	
	5.5	writing of management report
		GO TO 5.17
	END	
	5.6	definition of new specifications
	5.7	identification of the appropriate intervention type and planning
CASE	maintenance:	**GO TO** 5.8
	extension:	**GO TO** 5.10
END-CASE		

Table 6-a. Task structure of the KLIC 'maintenance and extension' phase: the first three steps [Guida and Tasso 94].

- **MAINTENANCE**

	5.8	revision of the knowledge base
IF	revision of target system is necessary	
THEN		**GO TO** 5.10
	5.9	writing of maintenance report
		GO TO 5.12

- **EXTENSION**

	5.10	revision of the target system and of the maintenance support system
	5.11	writing of extension report

- **RELEASE**

	5.12	installation, field testing, and refinement
	5.13	updating of manuals
	5.14	training of users and of the maintenance team
	5.15	certification and acceptance
	5.16	release

- **INTERVENTION CLOSING**

	5.17	updating of KBS history
	5.18	KBS development planning

UNTIL   the KBS is in use

Table 6-b. Task structure of the KLIC 'maintenance and extension' phase: the final four steps [Guida and Tasso 94].

A detailed description of all the above KLIC phases can be found in [Guida and Tasso 94].

4. Towards KBS Development Methodologies

A life-cycle model provides a general and abstract framework for organizing the KBS development processes. Therefore a life cycle model must not be too detailed, nor must deal with specific implementation aspects. As a consequence, at a lower level of abstraction, the implementation of the processes specified in a KBS life cycle, must be supported by specific methods and practices to be exploited for their execution. Moreover, the life cycle processes (the so called *primary processes*), must be sustained by two other categories of processes: *supporting processes*, i.e. those processes devoted to support primary processes by contributing to their success, such as for example, quality control and documentation, and management processes, i.e. those processes devoted to the basic activities of project management, resource allocation, infrastructuring, and so on.

Also the execution of supporting and management processes is performed by means of specific techniques, prescribing in detail how all the activities have actually to be carried out by their executors.

The comprehensive term *KBS methodology* refers to an integrated set of methods, techniques, and practices to effectively perform all the individual tasks involved in primary, supporting, and management processes. More precisely, it specifies, in detail, how each task should be carried out [Boehm 88]. A KBS methodology cannot be general since it is closely bound to the individual choices of a specific production or application environment, and it must therefore be developed as the result of the experience gained in a specific environment through several development projects.

A detailed definition and a comprehensive illustration of KBS supporting and management processes can be found in [Guida and Tasso 94].

REFERENCES

[Addis 85] T.R. Addis, *Designing Knowledge-Based Systems*, Kogan Page, London, UK, 1985.

[Banerji 90] R.B. Banerji (Ed.), *Formal Techniques in Artificial Intelligence - A Sourcebook*, Elsevier, Amsterdam, NL, 1990.

[Barr et al. 81-89] A. Barr, P.R. Cohen, and E.A. Feigenbaum (Eds.), *The Handbook of Artificial Intelligence*, Vol. I-IV, Addison-Wesley, Reading, MA, 1981-1989.

[Bench-Capon 90] T.J.M. Bench-Capon, *Knowledge Representation - An Approach to Artificial Intelligence*, Academic Press, London, UK, 1990.

[Boden 87] M.A. Boden, *Artificial Intelligence and the Natural Man* (2nd edition), MIT Press, Cambridge, MA, 1987.

[Boehm 88] B.W. Boehm, A spiral model of software development and enhancement, *Computer* 21(5), 1988, 61-72.

[Brachman and Levesque 85] R.J. Brachman and H.J. Levesque (Eds.), *Readings in Knowledge Representation*, Morgan Kaufmann, Los Altos, CA, 1985.

[Bundy 80] A. Bundy, *Artificial Intelligence: An Introductory Course*. Edinburgh University Press, Edinburgh, UK, 1980.

[Charniak and McDermott 85] E. Charniak and D. McDermott, *Introduction to Artificial Intelligence*, Addison-Wesley, Reading, MA, 1985.

[Charniak et al. 87] E. Charniak, C.K. Riesbeck, and J.R. Meehan, *Artificial Intelligence Programming* (2nd ed.), Lawrence Erlbaum, Hillsdale, NJ, 1987.

[Chorafas 90] D.N. Chorafas, *Knowledge Engineering*, Van Nostrand Reinhold, New York, NY, 1990.

[Davis and King 77] R. Davis, J. King, An Overview of Production Systems. In E.W. Elcock and D. Michie (Eds.) *Machine Intelligence 8*, Edinburgh University Press, Edinburgh, UK, 1977.

[Davis et al. 93] R. Davis, H. Shrobe, and P. Szolovits, What is knowledge representation? *AI Magazine* 14(1), 1993, 17-33.

[Edwards 91] J.S. Edwards, *Building Knowledge-Based Systems - Towards a Methodology*, Pitman, London, UK, 1991.

[Findler 79] N.V. Findler (Ed.), *Associative Networks: Representation and Use of Knowledge by Computers*. Academic Press, New York, 1979.

[Ford 87] N. Ford, *How Machines Think*, John Wiley & Sons, Chichester, UK, 1987.

[Genesereth and Nilsson 87] M.R. Genesereth and N.J. Nilsson, *Logical Foundations of Artificial Intelligence*, Morgan Kaufmann, Los Altos, CA, 1987.

[Guida and Tasso 89] G. Guida and C. Tasso (Eds.), *Topics in Expert System Design, Methodologies and Tools*, North-Holland, Amsterdam, NL, 1989.

[Guida and Tasso 94] G. Guida, C. Tasso, *Design and Development of Knowledge-Based Systems: From Life Cycle to Methodology*, John Wiley & Sons, Chichester, UK, 1994.

[Harmon and Sawyer 90] P. Harmon and B. Sawyer, *Creating Expert Systems for Business and Industry*, John Wiley & Sons, New York, 1990.

[Hayes-Roth et al. 83] F. Hayes-Roth, D.A. Waterman, and D.B. Lenat (Eds.), *Building Expert Systems*, Addison-Wesley, Reading, MA, 1983.

[Hilal and Soltan 93] D.K. Hilal and H. Soltan, Towards a comprehensive methodology for KBS development, *Expert Systems* 10(2), 1993, 75-91.

[Jackson 90] P. Jackson, *Introduction to Expert Systems* (2nd edition), Addison-Wesley, Reading, MA, 1990.

[Jameson et al. 97] A. Jameson, C. Paris, C. Tasso (Eds.), *User Modeling - Proceedings of the 6th Intl. Conference on User Modeling UM97*, Springer Verlag, Wien-NewYork, 1997.

[Laubsch 84] J. Laubsch, Advanced LISP programming, in T. O'Shea and M. Eisenstadt (Eds.), *Artificial Intelligence - Tools, Techniques, and Applications*, Harper & Row, New York, NY, 1984, 63-109.

[Minsky 75] M. Minsky, A Framework for Representing Knowledge. In P.H. Winston (Ed.) *The Psychology of Computer Vision*, McGraw-Hill, New York, NY, 1975.

[Nilsson 80] N.J. Nilsson, *Principles of Artificial Intelligence*, Tioga, Palo Alto, CA, 1980.

[Pearl 84] J. Pearl, *Heuristics*, Addison-Wesley, Reading, MA, 1984.

[Prerau 90] D.S. Prerau, *Developing and Managing Expert Systems - Proven Techniques for Business and Industry*, Addison-Wesley, Reading, MA, 1990.

[Rich and Knight 91] E. Rich and K. Knight, *Artificial Intelligence* (2nd edition), McGraw-Hill, New York, NY, 1991.

[Ringland and Duce 84] G.A. Ringland and D.A. Duce, *Approaches to Knowledge Representation: An Introduction*, John Wiley & Sons, New York, NY, 1984.

[Rolston 88] D.W. Rolston, *Principles of Artificial Intelligence and Expert Systems Development*, McGraw-Hill, New York, NY, 1988.

[Schalkoff 90] R.J. Schalkoff, *Artificial Intelligence: An Engineering Approach*, McGraw-Hill, New York, NY.

[Shapiro 87] S.C. Shapiro (Ed.), *The Encyclopedia of Artificial Intelligence*, John Wiley, New York, NY, 1987.

[Sowa 84] J.F. Sowa, *Conceptual Structures - Information Processing in Mind and Machine*, Addison-Wesley, Reading, MA, 1984.

[Waterman 86] D.A. Waterman, *A Guide to Expert Systems*, Addison-Wesley, Reading, MA, 1986.

[Winston 92] P.H. Winston, *Artificial Intelligence* (3rd edition), Addison-Wesley, Reading, MA, 1992.

EFFECTIVE KNOWLEDGE ELICITATION

P.W.H. Chung
Loughborough University of Technology, Loughborough, UK

Abstract

The construction of a knowledge-based system is an attempt to embody the knowledge of a particular expert, or experts, within a computer program. The knowledge used in solving problems must be elicited from the expert so that it can be acquired by the system. It has long been recognised that the elicitation of knowledge from the experts is a potential bottleneck in the construction of knowledge-based systems. This paper discusses some of the proven knowledge elicitation techniques and provides practical guidelines on how to apply them effectively. The techniques are described in the context of a single knowledge engineer acquiring knowledge from a single expert. This is the arrangement that most workers recommend; issues relating to multiple experts and multiple engineers are discussed in the final section.

1. Introduction

The construction of a knowledge-based system (KBS) is an attempt to embody the knowledge of a particular expert, or experts, within a computer program. The knowledge used in solving problems must be elicited from the expert so that it can be acquired by the KBS. A distinction between *knowledge elicitation* and *knowledge acquisition* is often made. Knowledge elicitation is the process of interacting with domain experts using techniques to stimulate the expression of "expertise"; knowledge acquisition is the process of extracting, transforming, and transferring expertise from a knowledge source to a computer program (McGraw and Harbison-Briggs, 1989). This paper deals primarily with the former. It has long been recognised that the elicitation of knowledge from the experts is a potential bottleneck in the construction of KBSs (Feigenbaum, 1977). The main reason for this, in many cases, is that experts find it hard to articulate and make explicit the knowledge they possess and use. Although experts perform their tasks skilfully and efficiently, they lose awareness of what they know. This is a common phenomenon in skilled behaviour (Welford, 1968). The expert may be unable to verbalise the skill, or if she attempts to do so, the skill goes to pieces. Johnson (1983) has called this "the paradox of expertise". The implication is that a knowledge engineer has to be skilled in helping experts to identify and formalise the domain concepts.

No doubt, even within a single domain, there are different types of knowledge. However, it is not clear how knowledge should be classified into different types. "Finding a way to taxonomise knowledge on a principled basis is a difficult and ambitious task that has eluded philosophers for thousands of years". (Gammack and Young, 1985). For the practical purpose of building KBSs, knowledge is normally divided into four types: facts, conceptual structures, procedures and rules. Facts are simply a glossary of terms and a list of domain entities. In an engineering domain, this type of knowledge may be a collection of engineering concepts and the names of the components of a structure or plant. The second type of knowledge, conceptual structures, describes the relationships between identified concepts and components. A procedure is simply a sequence of instructions on how to achieve a particular goal. Finally, rules are the reasoning part of the knowledge. They specify the relationships between situations and actions. A distinction between domain rules and meta rules (or strategic rules) is often made (Hayes-Roth et al, 1983). Domain rules deals directly with the concepts found in the domain. Meta rules provides strategic guidance in using domain rules.

Although there are many knowledge elicitation techniques proposed by researchers, only a few of them are used for building practical systems. Cullen and Bryman (1988) analysed 70 operational expert systems in Britain and the USA, and found that 90% of knowledge acquisition involved just four techniques: interviewing, documentation analysis, prototyping and case study analysis. Other researchers (Miles and Moore, 1989), and the present author, have also found card sorting a very useful technique. Another point from the paper by Cullen and Bryman is that the majority of cases used a combination of techniques, rather than just one, for the development of the systems. This is an important point because different techniques should be seen to be complementary. An experienced knowledge engineer would use whatever technique is appropriate during different phases of the project. Some studies have been carried out to match knowledge elicitation techniques with types of knowledge (for example see Gammack and Young, 1985; Welbank, 1987).

This paper discusses the above mentioned techniques and provides practical guidelines on how to apply them effectively. The techniques will be described in the context of a single knowledge engineer acquiring knowledge from a single expert. This is the arrangement that most workers recommend (for example see Hayes-Roth et al, 1983); issues relating to multiple experts and multiple engineers will be discussed in the final section.

2. Getting Ready

Assuming that a suitable application has been identified and the project proposal has been approved, both the knowledge engineer and the domain expert have to go through an initial phase of familiarisation before formal knowledge elicitation sessions begin. The start-up phase of a KBS development project has significant impact on later knowledge elicitation effectiveness.

2.1 Orientation of Expert

Experts who have been selected to provide the expertise for the development of a KBS are unlikely to have experienced similar projects before. They need to be adequately informed of what their involvement will be, what is expected of them, what the development goals are. Some authors (e.g. McGraw and Harbison-Briggs, 1989) recommend giving the experts an introduction to KBSs, although the author's own experience is that this is not necessary, and is sometimes counter productive (Chung and Kumar, 1987). Many experts are not interested in knowing how computer programs work. Furthermore, they may be sceptical about the various knowledge representation techniques. They are unlikely to be impressed, or convinced, by being told how their thinking process will be mechanised. Therefore, it is much more useful to concentrate on preparing them psychologically by describing to them what they should be expecting from the different knowledge elicitation sessions.

2.2 Orientation of Knowledge Engineer

The knowledge engineer has to be given time to prepare for the new project. She must be given the opportunity to meet the expert on a fairly informal basis. She must also be given time to carry out a number of preparatory tasks: to become familiar with the new domain, to set up a framework for gathering and recording information and to develop a knowledge elicitation plan.

The knowledge engineer should be provided with the background information to the project. She should also spend time reading up introductory and training material in the domain. This has several advantages. It will give her an overall view of the domain, some ideas on what needs to be found out, and to start a system for storing information. Once a project gets under way it will generate a lot of material: reports, tapes, transcripts of sessions, glossaries of terms, diagrams, etc. It is crucial that a record keeping procedure is set up before masses of information is generated. McGraw and Harbison-Briggs (1989) provide useful templates for knowledge acquisition, rule header and content forms.

On the practical side, a room suitable for knowledge elicitation purposes has to be found, either at the expert's site or at the knowledge engineer's. The room should be free from noise or other distractions. It is important that audio recording equipment, or video equipment if necessary, can be conveniently arranged in the room. A flip chart, or something similar, must be readily available.

2.3 Developing a Plan

Knowledge elicitation is a process that involves many activities and has different, though not well defined, stages. Developing a plan helps the knowledge engineer to set clear objectives and allocate resources. The plan also helps her to communicate with others about what she intends to do. As the project moves on, the plan is crucial for evaluating progress, and should be modified as progress is made and more is learned about the domain.

Initially, the knowledge engineer must try to scale the project and decompose the domain into major sections. She could do this by reading published literature, identifying how long it takes an expert to complete various tasks, and estimating the number of concepts in the domain. Once the size of the project has been scaled and the domain decomposed, the knowledge engineer then needs to decide whether to develop the components in parallel or in sequence.

When scheduling the knowledge elicitation sessions, the knowledge engineer needs to consider the availability of the expert and to allow adequate time for analysing the results of one session before starting the next — typically one to two days. If rapid prototyping is to proceed in parallel, time has to be allowed for programming and documenting. When to begin prototyping during the project is an important decision (see the section on rapid prototyping later).

3. Knowledge Elicitation Techniques

3.1 Interview

Direct interviewing is the technique most familiar to experts and knowledge engineers. It is good to start the knowledge elicitation process using a technique that the expert is comfortable with. An interview may range from an informal chat to a highly focused and structured question-and-answer session. Researchers have pointed out a number of drawbacks with interviewing, particularly the unstructured type (McGraw and Harbison-Briggs, 1989; Musen, 1989):

1) an expert may perceive the lack of structure in an interview as requiring little preparation on her part prior to the interview;
2) data gathered from an unstructured interview is often unrelated, exists at varying levels of complexity, and is difficult to analyse
3) an expert's response to a question may depend in a subtle way on how the question is asked;
4) an expert may volunteer plausible answers that may not reflect her true behaviour when asked questions about tacit processes.

Although useful information may not be gathered in a single interview, the drawbacks mentioned can be overcome by a series of interviews. As long as the knowledge engineer is aware of these drawbacks, interviewing remains a popular and productive technique.

The knowledge engineer must set clear objectives for each interview, and the expert needs to be told them at the start of the session, or well in advance if preparation is required. Each session should be tape recorded, and to get the most out of it the recording should be transcribed and analysed before the next session. A one hour session with an expert will normally take three hours to transcribe.

There is an important difference between interviewing for clarification and interviewing for content. It is good practice is to divide a session into two parts, one for reviewing the previous session and one for exploring new grounds. The reviewing part will help the knowledge engineer to clarify her own understanding. She should have prepared diagrams, tables, etc, to show to the expert for comments. The need for some sort of formal representation for communicating with the expert is stressed by Johnson (1987). When using diagrams in an interview, it is essential that they are properly labelled. The conversation should refer to the diagrams explicitly by their labels. Otherwise, it would be very difficult to make sense of the transcript.

3.2 Document Analysis

By analysing documents, the knowledge engineer can obtain a lot of useful information without the need for direct interaction with an expert. For example, almost half of the rules in the prototype KBS for the design of industrial buildings described by Chung and Kumar (1987) were gleaned from the literature. The main advantage of document analysis is that it allows the knowledge engineer to initially structure and define the knowledge base without using much of the expert's time. One problem with indiscriminate trawl for rules in the literature is that paper may be out of date. Another is that a rule may belong to a given set used by one designer. It may not be consistent with some other rule taken from a different design philosophy.

Useful documents come in many different guises, and it is important that the knowledge engineer obtain a list of reference material that is used by, or produced by, the expert early on in the project. Whatever information that the knowledge engineer obtains from the documents needs to be fedback to the expert for verification.

3.3 Card Sort

Experts use specialist knowledge to solve problems; they are also likely to have a global perspective on how a domain is organised. Card sort is appropriate where there is a large set of concepts which need to be organised into a manageable form. The basic procedure is similar to the categorical knowledge elicitation technique described by Regan (1987):

1) identify a set of domain concepts, either from the literature, an introductory talk or the domain expert;
2) before the card sort session, write each concept on a small card in advance;
3) ask the expert to sort the cards into groups;
4) ask the expert to label each group;
5) discuss with the expert about each group to determine its characteristics;
6) ask the expert to specify the relationship between the groups and to organise them into a hierarchy.

Miles and Moore (1989) reported very favourable results using this technique in the conceptual bridge design domain. It was originally thought that the experts would find it an easy exercise. "However, on implementing the card sort technique, it became apparent that the experts found that the exercise tested their resourcefulness, and forced them to consider problems in a new way. The card sort proved both popular and successful."

3.4 Rapid Prototyping

Building a KBS is an iterative process. Unlike conventional system development projects, most, if not all, KBS projects start with a fairly ill defined specification. The specification and the system are then successively refined as the knowledge engineer finds out more about the domain. Using the rapid prototyping technique, the expert is confronted with the behaviour of an unfinished version of the system which is modified in the light of her comments. Each iteration brings the behaviour of the system closer to completion although, since it is often carried out without a clear defined notion of completion, it is perhaps better thought of as iteration towards adequate achievement.

Some people argue that one should start building the prototype version of the KBS as soon as the process for solving a single example is felt to be well understood. This emphasises the importance of embodying what the expert says in a computer program as quickly as possible. Showing a working system to an expert has two advantages. One is to get feedback on how to improve the system. The other is to keep the expert interested.

However, there are a number of problems with getting involved in rapid-prototyping very early on in the project (Inder et al, 1987). Without developing a good understanding of the application, inappropriate knowledge representation techniques and reasoning strategies may be chosen for the implementation. Focusing on implementation may also lead to the neglect of documentation. While some experts may be enthusiastic about a partially running system, others may be put off by it because it is so trivial and crude. Therefore, the timing of the development of a prototype and the purpose of the system have to be clearly established.

If the prototype is to be developed into the final product then prototyping should wait till much of the domain information is formalised. However, where the prototype is to be used as a tool for testing ideas, studying feasibility or validating requirements and specification, it is useful to have knowledge elicitation and prototyping proceed in parallel. In this latter case, once the lessons from the prototyping exercise have been learned the system is scrapped and a new one is designed and built based on the experience gained. This fits in well with Frederick Brooks Jr's idea of "plan to throw one away" (Brooks, 1979) for conventional software projects. He argues that "in most projects, the first system built is barely usable. It may be too slow, too big, awkward to use, or all three. There is no alternative but to start again, smarting but smarter, and build a redesigned version in which these problems are solved. .... all large-system experience shows that it will be done. ..... The management question, therefore, is not *whether* to build a pilot system and throw it away. You *will* do that. The only question is whether to plan in advance to build a throwaway, or to promise to deliver the throwaway to customers."

3.5 Case Studies

Case study is a general term for a set of techniques for probing an expert's "train of thought" while she solves a problem. Another term for case study is *process tracing* (Waldron, 1985). A representative problem or case is given to the expert. She is instructed to solve the problem or to make a decision in the usual way. The expert's behaviour is recorded as she works through the problem. The protocol is then transcribed and analysed. In this way, the knowledge engineer is given not only the answer to the problem but also the information about the problem-solving process itself. In practice this way of working is found to be very helpful. Though experts may have difficulty in stating the general rules that they use, they can usually identify the specific rules that they are applying. However, it is easy for familiar ideas to be taken for granted, so experts need to be kept aware of any tendency towards omitting "trivial" details. To apply the case study method effectively a set of representative problems has to be chosen and used, otherwise there could be serious errors of omission.

There are three different ways of generating protocols:

1) think-aloud protocols - the expert thinks aloud during the solving of a problem;
2) retrospective verbalization - the expert completely solves a problem before reporting how it was solved.
3) discussion protocols - a small number of experts discuss with one another as they attempt to solve a problem.

These variations each have their own advantages and disadvantages. The major problem with think-aloud protocols is that the reporting may interfere with the expert's task performance. Related to this is any need to conform to real time constraints. For example, solving a maths problem allows the mathematician to stop and ponder. However, an operator dealing with an emergency situation may require immediate responses. Another type of constraint is that it may not be appropriate for an expert working in a face-to-face situation to discuss her thought process. For example, a doctor seeing a patient, or a negotiator trying to secure a contract. These criteria may help when having to decide between think-aloud protocols and retrospective verbalization.

Expert system projects are often based on collaboration with a single expert. However, discussion protocols are helpful because they provide different perspective on how a problem may be solved by clarifying alternatives and resolving conflicts. The problem here is that of managing the discussion. Avoiding the problem, the strategy that Mittal and Dym (1985) adopted was to interview one expert at a time. Although this arrangement worked for them, it provides very little opportunity for the experts to interact with one another and to discuss issues.

4. Team Effort

4.1 Multiple Experts

It is widely recognised that involving more than one expert in a project can be problematic. The experts may not get on and they may disagree on many issues. Personality problems with experts can be avoided by careful selection. The purpose is to form a *team* of experts, not just a collection of individuals. The fact that experts disagree may not necessarily be a bad thing. Sometimes, rather than a single view, it is desirable for a KBS to have a consensus view, or for it to be able to highlight alternative views and to give reasons that support them. Another reason why more than one expert is needed is that some projects may require several specialist experts.

As mentioned above, when multiple experts are used in a project, they can be consulted either as a group or individually. Individual consultation has the advantage that the discussion is easier to manage, but it does take more of the experts' time if a number of them are to be asked the same questions. Another disadvantage of individual consultation is that some experts may not like the knowledge engineer to reveal what has been said to other experts without obtaining prior permission. This could be due to a number of reasons: confidentiality, unwillingness to expose one's knowledge for others to assess, or fear of being mis-represented. For whatever reason, the knowledge engineer has to spend additional time to go over with the expert what she intends to reveal to others. The need to discuss what one expert said with another is less serious if multiple experts are being used because they are specialists of different subjects.

In a group discussion different views will be proposed and discussed. It is therefore a very effective way of discovering alternatives and resolving conflicts. However, it is important to avoid the situation where a single expert dominates the conversation, which can occur if one expert is more senior than the others. This possibility again emphasises the importance of selecting the right team of experts.

4.2 Multiple Knowledge Engineers

Knowledge engineers do not always work alone. There are many large KBS development projects that involve a team of knowledge engineers. When two or more people are working on a software project the need for rigorous management is obvious. While many project management issues - like coding standards, documentation, system integration, system modification - are in common with the development of conventional systems, there are a few that are peculiar to KBSs. In particular, these issues relate to knowledge elicitation.

After going through a few interviewing sessions together a rapport will have been built up between the expert and the knowledge engineer. Therefore, to take advantage of this, throughout a project, or as far as possible, each expert should be interviewed by the same knowledge engineer. This arrangement will also avoid the situation where another knowledge engineer may frustrate the expert by not knowing, or not understanding, what has already been done in the previous sessions. If a knowledge engineer has to be replaced by another then the new person should sit in two or three sessions before taking over.

Although audio recording of a session provides an exact record of what went on in a session, the information on a tape is not very accessible. One needs to know what is on the tape and needs to search for the particular piece of information sequentially, which can be very time consuming. On a project that involves a team it is important that the elicited knowledge should be made easily accessible to all team members. Therefore, there is the added emphasis on tape transcription. An index, or a summary, of the content of a tape is far from adequate. Tapes must be transcribed in full. Otherwise, there will be a lot of wasted effort.

Acknowledgements

The author is grateful to British Gas and The Fellowship of Engineering for financial support through a Senior Research Fellowship. He would also like to thank Dr Robert Inder, University of Edinburgh, and Professor Frank Lees for providing helpful comments on a draft of the paper.

References

Brooks, F.P. Jr. (1979) The Mythical Man-Month: Essays on Software Engineering. Addison Wesley.

Chung, P.W.H., and B. Kumar (1987) Knowledge Elicitation Methods: A Case-Study in Structural Design. The Application of Artificial Intelligence Techniques to Civil and Structural Engineering (Ed. B.H.V. Topping), Civil-Comp Press, Edinburgh.

Cullen J. and A. Bryman (1988) The Knowledge Acquisition Bottleneck: Time for Reassessment? Expert Systems, Vol 5(3), August, 1988.

Feigenbaum, E. (1977) The Art of Artificial Intelligence: Themes and Cases Studies of Knowledge Engineering. IJCAI-77, 1014-1029.

Gammack, J.G. and R.M. Young (1985) Psychology Techniques for Eliciting Expert Knowledge. In Research and Development in Expert Systems (Ed. M.A. Bramer). Cambridge University Press.

Hayes-Roth, F., D.A. Waterman, D.B. Lenat (Eds.) (1983) Building Expert Systems. Addison-Wesley.

Inder, R., P.W.H. Chung and J. Fraser (1987) Towards a Methodology for Incremental System Development. Proceedings of Workshop on Knowledge Acquisition for Engineering Applications (Eds. C.J. Pavelin and M.D. Wilson). Report No. RAL-87-055. Rutherford Appleton Laboratory.

Johnson, P.E. (1983) What Kind of Expert Should a System Be? Journal of Medicine and Philosophy, 8:77-97.

Johnson, N.E. (1987) Mediating representations in knowledge elicitation. Proceedings of the First European Conference in Knowledge Acquisition for Knowledge Based Systems. University of Reading.

McGraw, K. and K. Harbison-Briggs (1989) Knowledge Acquisition: Principles and Guidelines. Prentice-Hall.

Miles, J.C. and C.J. Moore (1989) An Expert System for the Conceptual Design of Bridges. The Application of Artificial Intelligence Techniques to Civil and Structural Engineering (Ed. B.H.V. Topping), Civil-Comp Press.

Mittal, S. and Dym, C.L. (1985) Knowledge Acquisition From Multiple Experts. AI Magazine, pp32-36, Summer 1985.

Regan, J.E. (1987) A Technique for Eliciting Categorical Knowledge for an Expert System. Proceedings of AAAI-87.

Waldron, V. (1985) Process tracing as a means of collecting knowledge for expert systems. Texas Instruments Engineering Journal, 2:6, pp90-93.

Welbank, M.A. (1987) Perspectives on Knowledge Acquisition. Proceedings of Workshop on Knowledge Acquisition for Engineering Applications (Eds. C.J. Pavelin and M.D. Wilson). Report No. RAL-87-055. Rutherford Appleton Laboratory.

Welford, A.T. (1968) Fundamentals of Skill, Methuen, London.

MODELLING MECHANICAL BEHAVIOUR WITHOUT MECHANICS

J. Bento
Technical University of Lisbon, Lisbon, Portugal

1. Introduction

The modelling of mechanical behaviour of structures or, simply, that of solids bodies, has undergone a process of enormous maturation through the history of Mechanics, in the last two centuries. Depending on the then existing scientific paradigms, each of the steps of improvement in the modelling of mechanical behaviour of solid bodies has taken various forms; however, regardless of using a more rational approach or one of, predominantly, an empirical nature, *mechanics* has been invoked as the obvious supporting discipline for the analysis and synthesis of behaviour. Hence, the immensely rich spectrum of modelling attitudes spanning from experimental, through purely theoretical to computational methods.

In perhaps too simplistic terms, one could describe all these modelling approaches – either theoretical, experimental or computational – as *physical* in a broad sense, given the attempt they all encompass of modelling a (physical) reality by describing the associated (physical) phenomena using a set of representational tools of a physical nature. In the case of structures, the emphasis has been in the mimicking of the phenomena itself through their governing mechanical laws.

The main objective of the present work is to illustrate that, under specific conditions and motivations, it is possible to adopt a modelling attitude that, when compared with the previous ones, may be described as *non-physical* while still showing a high level of rationality, for it totally neglects the physical mimicking of the phenomena associated to mechanical behaviour.

The tool that enables such an approach – artificial neural networks – is able of levels of computational efficiency in representing mechanical behaviour that, in some cases, clearly overwhelms the traditional ones provided by computational or theoretical mechanics.

Along this chapter, very basic principles of artificial neural networks are initially introduced; next, some models and architectures are presented and the illustration of their application in the modelling of mechanical behaviour is provided through a number of examples.

The chapter closes with a discussion on the appropriateness of this approach, namely in terms of opportunity in comparison with other approaches to modelling of mechanical behaviour.

2. Basic concepts

Artificial neural networks and, more generally, "connectionist approaches" to Artificial Intelligence (AI), share few common methods with the branch of AI denoted as *knowledge based systems* – the basic theme covered by this book –. While the former enable no explicit representation of the knowledge they actually incorporate, the latter represent explicitly, in a symbolic manner, expert knowledge acquired elsewhere, thus their affiliation to the "symbolic approach" to AI.

2.1 Empirical approaches to neural networks

The original inspiration to the development of artificial neural networks, which naturally has influenced their name, was of an empirical nature and originated in neuroscience. Indeed, the reference that unanimously has founded the discipline by suggesting a simple model for the computational operation of neurons is due to McCulloch and Pitts (1943), a neuro-physiologist and a mathematician, respectively. This model proposes a computational scheme similar, though very simplified, to that of a biological neuron.

That operation may be described in a simplified manner as follows: a neuron's *cell* receives and processes an inflow of electrochemical signals, incoming from *axons* of other neurons and filtered (reinforced or weakened) in the *synapses*; this signal, once processed by the cell, originates an electrical outflow that is directed at other cells within neurons at downstream of the network (Figure 1).

2.1.1 McCulloch and Pitts' artificial neurons

A large number of architectures for artificial neural networks has been proposed and developed. However, given the scope of the present text it would be inappropriate to discuss in great detail a large number of architectures, reason why a single class of networks is presented in this chapter – *unidirectional* (feedforward) networks – and, even so, at a very introductory level.

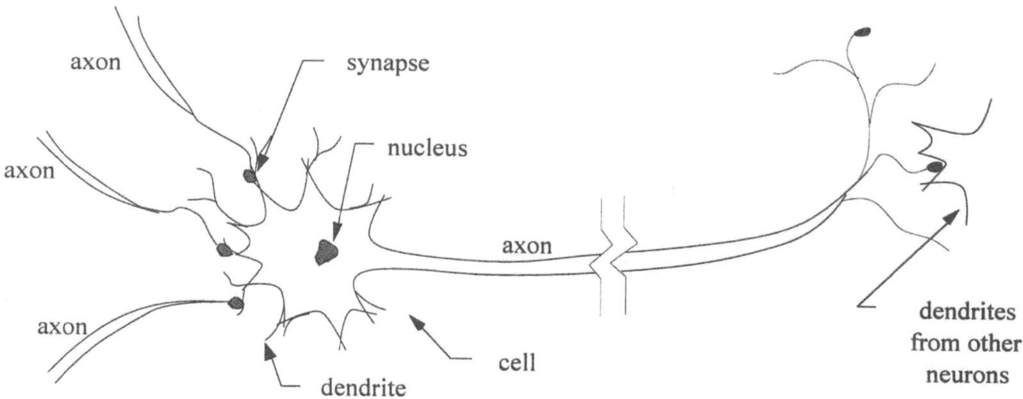

Figure 1 - Biological neuron

It is felt that the presentation of this single class of artificial neural networks is sufficient to illustrate the approach to modelling mechanical behaviour that is envisaged.

Figure 2 describes the first and one of the most famous of those types of neural models – that of McCulloch and Pitts (1943). In this model, the inflow taking place at the cell is represented by the input values – $x_i(t)$, produced in upstream neurons at time t – while a simplified scheme is proposed for representing the synaptic weight, w_i – positive weights are excitatory, negative ones are inhibitory and null ones denote the absence of synapse; all excitatory weights have the same value, the same occurring to inhibitory ones. Input values are weighted and summed (equation (1)) in order to set the total input to the so called *activation function* at time $t - g(t) -$.

The neuron's output at time $t+1$ is either 0 or 1 since it behaves as a binary processing

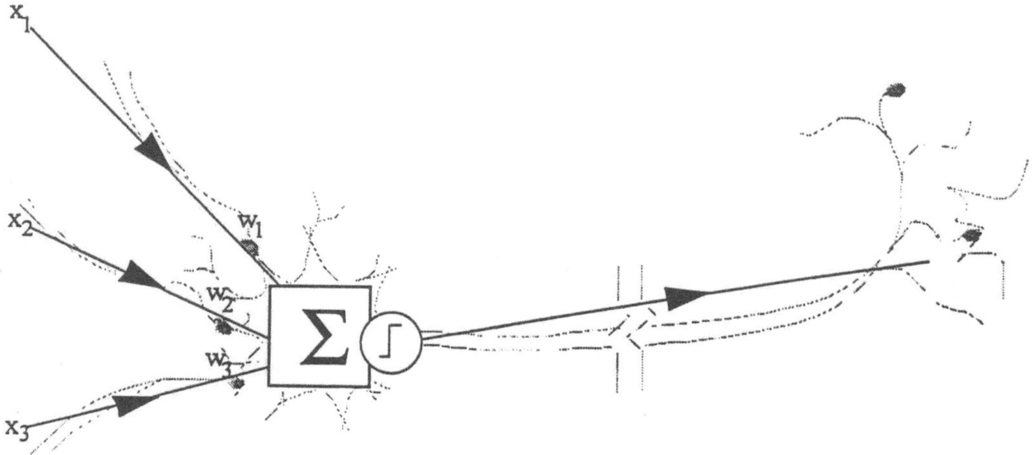

Figure 2 - McCulloch and Pitts artificial neuron

unit: if the sum stands below a given threshold $-\mu-$, the unit does not fire (it fires 0); likewise, if the argument exceeds the threshold, it fires 1, as established by the *output function* of equation (2) $-\Theta-$ and is illustrated in Figure 3.

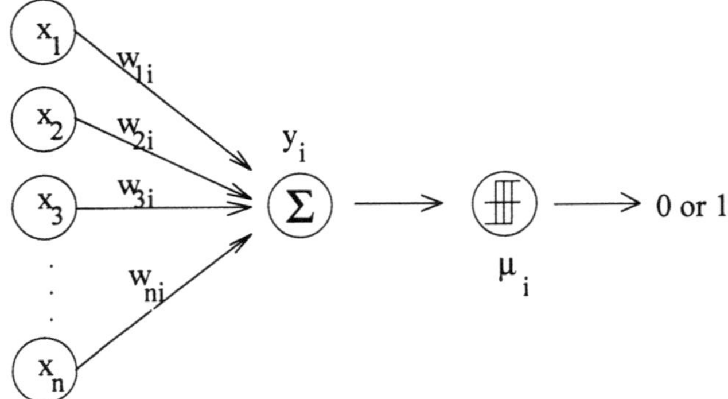

Figure 3 - Schematic view of a McCulloch and Pitts artificial neuron

$$g(t+1) = \left(\sum_{i=1}^{n} w_{ji} x_i(t) \right) \tag{1}$$

$$\Theta(t+1) = \begin{cases} 1 & g(t+1) \geq \mu_i \\ 0 & g(t+1) < \mu_i \end{cases} \tag{2}$$

Although simple, the McCulloch and Pitts artificial neuron, or some judicious assemblies of them, would be capable of, in principle, universal computation (Hertz, Krogh and Palmer, 1991), given the possibility of choosing suitable weights w_{ij}. The previous assertion obviously disregards computational efficiency.

The main problem associated to the generalisation of the use of artificial neural models, such as the ones from McCulloch and Pitts or their successors, was related to the difficulty in finding the appropriate weights w_{ij} that would enable the establishment of the modelling relations that the artificial neural systems were aimed at.

2.1.2 *Perceptrons*

In the early 1960s, a number of authors have independently introduced a new class of unidirectional artificial neural networks called *perceptrons* or *adalines*, the most relevant of which being Frank Rosenblatt (1957, 1958, 1962).

One of the foremost important contributions introduced in NN research by perceptrons was the demonstration that, for every problem that perceptrons could compute, it was possible to automate an iterative and convergent process of search for appropriate weights by means of a *learning algorithm* supported by the *perceptron convergence theorem* (Rosenblatt, 1962).

Figure 4 summarises one possible perceptron architecture. Input values are represented by x_i and weights are denoted by w_i. Like in the previous model, weighted inputs are summed by a *summation* function g (equation(3)) in order to set the total input to the *activation function*, Θ (equation (4)).

In the typical perceptron of Figure 4., the activation Θ is also defined as a binary function: for a sum g below a given threshold w_0, the unit does not fire (fires 0); likewise, if the threshold is exceeded, it fires 1.

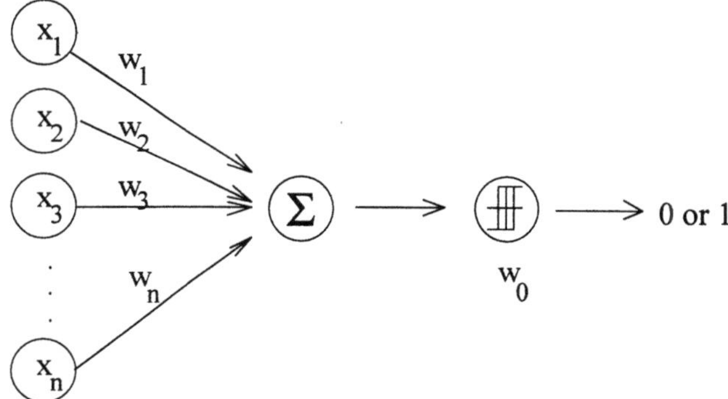

Figure 4 - Threshold in the activation function

$$g(x) = \left(\sum_{i=1}^{n} w_{ji} x_i \right) \tag{3}$$

$$\Theta(x) = \begin{cases} 1 & g(x) \geq w_0 \\ 0 & g(x) < w_0 \end{cases} \tag{4}$$

A more pragmatic treatment of the threshold concept, from an implementation point of view, suggests that it may be considered as an additional negative weight, associated to a neutral input (an input of 1), as depicted by Figure 5. In that case, equations (3) and (4) are to be replaced by equations (5) and (6).

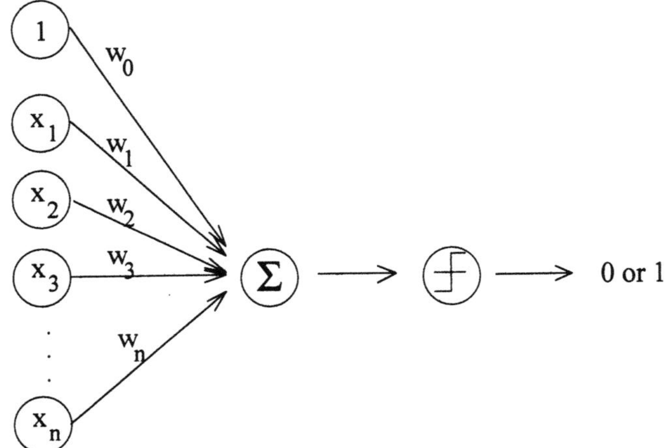

Figure 5 - Threshold as additional weight

$$g(x) = \left(\sum_{i=1}^{n} w_i x_i \right) - w_0 \qquad \text{or} \qquad g(x) = \left(\sum_{i=0}^{n} w_i x_i \right) \qquad (5)$$

$$\Theta(x) = \begin{cases} 1 & g(x) \geq 0 \\ 0 & g(x) < 0 \end{cases} \qquad (6)$$

In order to illustrate the functioning of single perceptrons, let us consider the following traditional example (a linearly separable classification task in a two dimensional space) as a means to identify a process of establishing appropriate weights:

Let a given perceptron be fed with pairs of co-ordinates (x_1, x_2) referring to <u>some</u> of the white and black dots illustrated in figure 3; let an output of 0 or 1 be associated to, respectively, white or black dots.

Such perceptron would have only 2 input units – one for receiving the values of x_1 and another one for those of x_2, as illustrated in Figure 6.

It is relatively easy to establish the weights that would enable the perceptron to differentiate between every white and black dots, generalising, thus, the ability to classify them. Indeed, considering that equation (5) can be rewritten for this specific problem as follows

$$g(x) = w_0 + w_1 x_1 + w_2 x_2 \qquad (7)$$

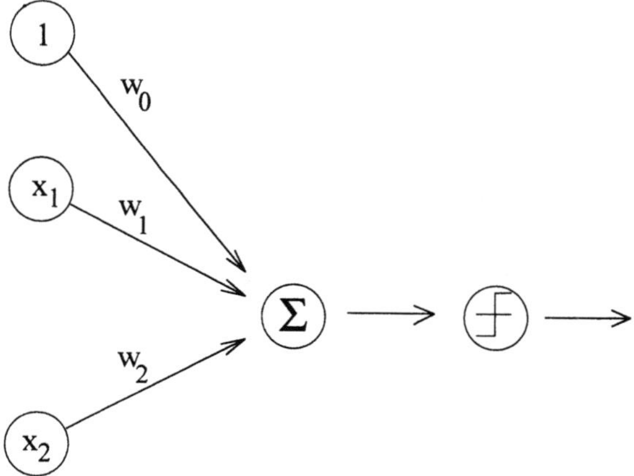

Figure 6 - Architecture for *white/black dots* problem

it results that, from equation (6), the perceptron's activation will change its result (from 0 to 1) for the combination of input values (x_1, x_2) that produce a sum of zero. At this stage, as pointed out in Rich and Knight (1991) "a slight change of inputs could cause the device to go either way", as clearly shown by equation (8).

$$0 = w_0 + w_1 \, x_1 + w_2 \, x_2 \qquad\qquad (8)$$

Equation (8) is, indeed, that of a line

$$x_2 = - \frac{w_1}{w_2} x_1 - \frac{w_0}{w_2} \qquad\qquad (9)$$

Such line, fully defined by the weights w_i, <u>separates</u> the pairs of input values (x_1, x_2) that would compel the perceptron to fire 0, from those that will produce 1.

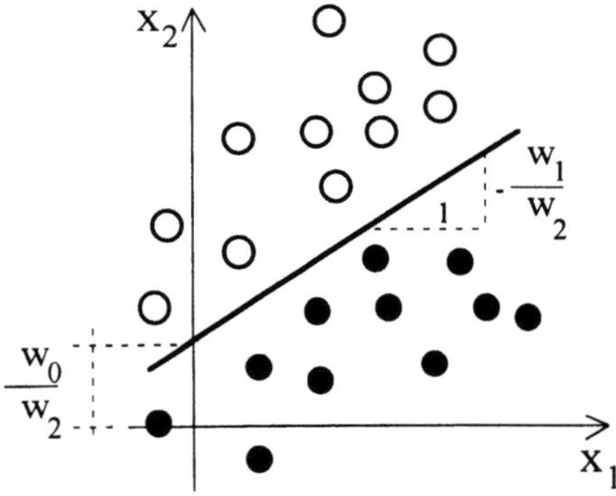

Figure 7 - Linearly separable problem

It would, therefore, be easy to compute, in a more or less iterative way, appropriate initial values of w_0, w_1 e w_2, by simply using 3 specific pairs of input/output vectors (3 known points), preferably some white and some black. For any additional given input pair that would make the perceptron to eventually compute a wrong output, one could correct each w_i proportionally to the associated input x_i.

The problem just described – that of linear separation in a two dimensional space – may be generalised to a similar problem of a larger dimension. The equivalent to the concept of a *separation line* would, in that case, be generalised to that of a *separation hyperplane*. Figure 8, adapted from Russel and Norvig (1995) illustrates the same problem in R^3, where the separation hyperplane is still an Euclidean plan.

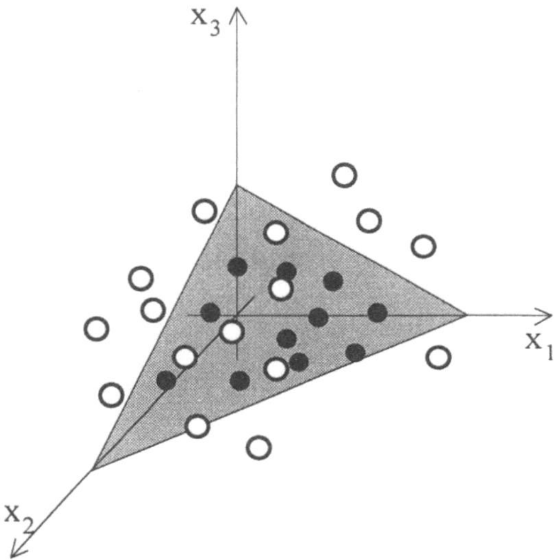

Figure 8 - Linearly separable problem in a three dimensional space

However, there is a different type of generalisation that cannot be achieved: the present classification problem, even in a bi-dimensional space, would not be generalisable to the problem represented by the sets of dots of Figure 9, since it clearly does not show linear separation between the two types of occurrences.

The introduction of perceptrons and that of a learning algorithm has, therefore, revealed itself of less interest than predicted, given the limitation of their computational capabilities being restricted to that of representing merely linearly separable functions. Such severe limitation was identified as the major impairment to the generalisation of the use of artificial neural networks, as put forward in the pessimistically seminal book by Minsky and Papert, *Perceptrons* (1969).

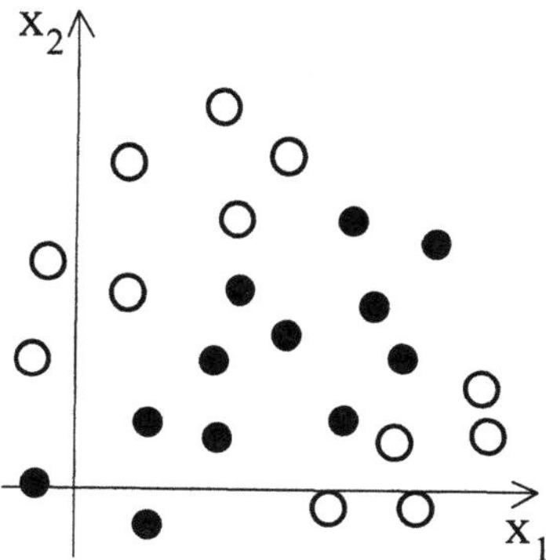

Figure 9 - Non-linearly separable problem

2.1.3 Multi-layered feedforward networks

In very broad terms, this limitation, the identification of which was responsible for a virtual halt in artificial neural networks research of many years, can be overcome by using networks of multiple layers of perceptrons (Figure 10).

The study of feedforward artificial neural networks formed by multiple layers of *units* (perceptrons or others) dates from the transition from the 1950s decade to the 1960s (Rosenblatt, 1957; Widrow and Hoff, 1962). The problem that persisted was, however, that, a higher level of complexity in the networks' architectures, such as would happen in networks with many layers of many units each, would complicate the crucial task of finding appropriate weights suitable for complex computations.

It was therefore only many years later that the introduction of such a learning algorithm

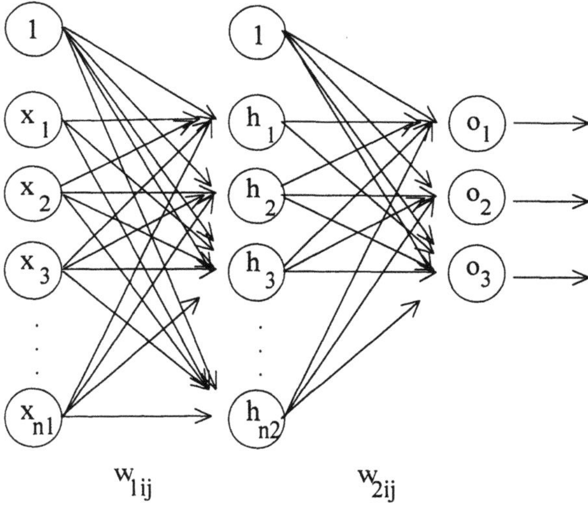

Figure 10 - Multi-layered feedforward network

enabled the emergence of artificial neural networks as a practical tool useful for computational tasks of higher complexity. This method – the algorithm of error backpropagation – was independently discovered by several authors (Bryson and Ho, 1969; Werbos, 1974; Parker, 1985 and LeCun, 1986) but was never put forward as a possible solution to the learning of weights. In 1986, Rumelhart, Hinton and Williams (1986a and 1986b) systematised, refined and publicised these achievements and a first version of a series of learning algorithms for multi-layered feedforward artificial neural networks, eventually became available.

Following that, and pursuing a paradigm of scientific maturation similar to the one that took place in many other areas of science and engineering(but mostly of engineering),the domain of neural computing was given unprecedented theoretical robustness in the forthcoming years: in 1988, Cybenko demonstrated that a network with only 2 intermediate layers is capable of any mapping between an input space R^n into an output space R^m. Such achievement was later refined by Hornik et al. (1989), by demonstrating that the computational universality of multi-layered networks would be guaranteed by a single intermediate layer, if formed by a sufficient number of (hidden) units.

2.1.4 Backpropagation networks

Feedforward multi-layered networks trained by error backpropagation are often called *backpropagation networks*. The algorithm of error backpropagation will be presented with reasonable detail in this chapter, given the crucial importance it assumed in the development of successful applications of artificial neural networks.

A basic difference between simple perceptrons and backpropagation networks is that, while the former sums the weighted inputs to produce a binary signal (0 or 1), the latter compute a real value (e.g. between 0 and 1). This is a feature of the backpropagation algorithm that requires an activation function that is differentiable, hence the need to replace the signal or step activation functions used by perceptrons by similar but continuous and differentiable ones.

Typically, backpropagation networks have activation function called *sigmoids*. Equation (10) represents a binary sigmoid function ranging within]0, 1[(Figure 11 a) while equation (11) denotes a bipolar sigmoid in the range]-1, 1[(Figure 11 b).

$$\Theta(x) = \frac{1}{1 + e^{-g(x)}} \qquad\qquad (10)$$

$$\Theta(x) = \frac{2}{1 + e^{-g(x)}} - 1 \qquad\qquad (11)$$

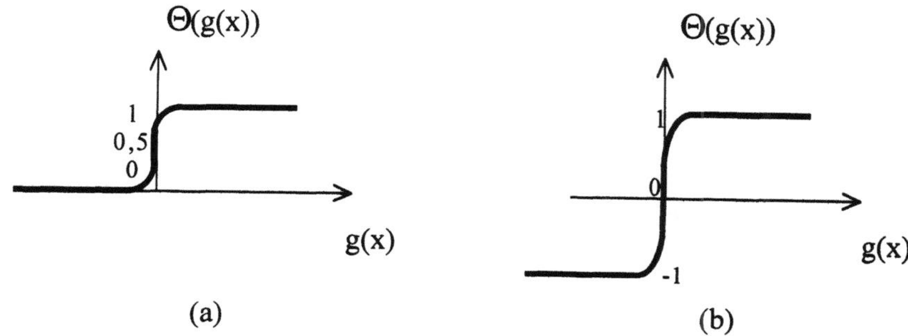

Figure 11 - Sigmoid functions

In both cases $g(x)$ remains the sum or summation function given by equation (5).

The training of backpropagation networks follows a three stage procedure: 1) feeding of the network with the input vectors (training); 2) calculation and backpropagation of the computed error and 3) adjustment of the weights in conformity. Learning through these stages evolves by following a principle already enforced when discussing the training of perceptrons: the weights should be corrected in the proportion of their contribution to the formation of the output.

The algorithm, fully generalisable for any number of layers, may be described using a 2 layer network (a network with 1 hidden layer) as described in pseudo-code at Table 1.

```
BPTraining(netTopology, SetOfExamples, α)
  w₁ᵢⱼ←rand(-0.1;0.1)              /* generate random weights */
  x₀←1.0                          /* initialise thresholds */
  h₀←1.0
  repeat
   for each x in SetOfExamples        /* going forward */
    do
    { hⱼ ←        1                 /* compute output of the 1ˢᵗ layer */
           ───────────────
                  -Σw₁ᵢⱼxᵢ
           1+e

     oⱼ ←        1                  /* compute total output (of the 2ⁿᵈ layer) */
           ───────────────
                  -Σw₂ᵢⱼhᵢ
           1+e

     δ₂ⱼ ←oⱼ(1-oⱼ)(yⱼ-oⱼ)  /* compute total error */

     δ₁ᵢ ←hⱼ(1-hⱼ)Σδ₂ᵢw₂ᵢⱼ  /* compute intermediate error */
                  i=1

                                    /* propagating backwards */
     Δw₂ᵢⱼ← α δ₂ⱼ hᵢ               /* adjust weights leading to output layer */
     Δw₁ᵢⱼ = α δ₁ⱼ xᵢ              /* adjust weights - intermediate layer */
    }
   until δ₁ᵢ < allowadbleError
  return trainedNet
```

Table 1 - Pseudo-code for backpropagation algorithm

Theoretically, the algorithm follows a gradient descent method employed to minimise a norm of the total network error (e.g. the square root of the sum of the square of the errors).

The algorithm suggests, therefore, a simple training strategy by which the network computes its output while working forward and propagates the eventually found errors, backwards. Every time that such a cycle is closed for the full set on input/output vectors used for training, an *epoch* is said to be finished. The number of training epochs a network needs to undertake is highly dependent on the quality of the subjected data, of the network topology and of many other factors. Typical training sessions can take in the order of thousands of epochs until a network stabilises in an acceptable error.

However, there are many problems and solutions associated with the implementation of the algorithm as described, namely the possibility that the search for this minimum stops in local minima, that would fall out of the scope of the present text. Introductory readings on the subject may be found, for example, in Hertz, Krogh and Palmer (1991), Rich and Knight (1991), Fausett (1994) and Russel and Norvig (1995), among others.

2.2 Generalisation

Once trained, backpropagation networks may be used for computing the output of unknown input vectors. The creation of that capacity, called *generalisation*, is indeed the ultimate goal of the training effort.

In order to evaluate the generalisation capacity of a trained network, it is usual to leave aside of the training process a number of input/output pairs that were initially available. Once convergent, the network is *tested* with those input/output vectors: when fed by test inputs, the network ought compute the associated outputs below an acceptable error.

A well known and interesting problem that is related to generalisation is that of overfitting, a problem also arising in statistical processes. To understand this phenomenon, let us consider Figure 12 that illustrates a typical plot of the evolution of a network error (in fact, the inverse of the total error) *vs.* the number of epochs, during training.

In zone ❶, one may notice an expected progression of the generalisation capacity of the network: the more training epochs, the lower the network error and the error associated with the classification of unknown (test) vectors.

Zone ❷ illustrates a halt in the learning process: despite the evolution of epochs, the modified weights confer no better discriminating capacity to the network.

As training proceeds, the previous phase is eventually overcome when the training error progresses again (zone ❸). However, the curve relating to the test set worsens significantly.

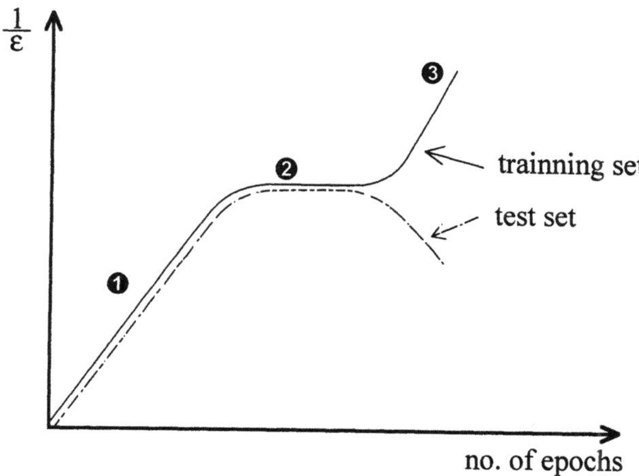

Figure 12 - Evolution of generalisation capacity with training

Here is where overfitting has taken place: the fact that the weights have produced lower training errors is associated to an over-adaptation to the training input/output pairs. Such over-adaptation obviously decreases the network capacity to correctly classifying unknown input vectors (those that have not been used during training).

Therefore, supervision of the learning phase could be of help, in the sense that the training should stop in the transition from zone ❶ to zone ❷.

3. Selected mechanical problems

Two very simple examples of mechanical problems to be modelled through artificial neural networks have been prepared: the first one attempts to identify cross sectional properties of steel and composite columns subjected to cyclic loading, from the analysis of hysteretic diagrams of their behaviour. The problem is illustrated in Figure 13, while Figure 14 provides several representations of typical hysteretic diagrams; the second example consists of a highly non-linear problem of frictional contact, generically described in Figure 15. In this case, a pin-on-flat system with a concentrated mass at a level above the contact surface is subjected to gravity forces and prescribed displacements which induce displacements and reactions; the calculation of these is, when possible, both very complex and time (CPU) consuming.

In the first problem, no analytical solution would enable appropriate modelling, although sufficient experimental data exists to enable training of some artificial neural network(s). Moreover, approximate solutions to this problem that can only be obtained by numerical approaches are extremely CPU intensive to the point of rendering the use of such models unbearable for practical purposes; hence, the usefulness of the present approach, where the network is trained with results obtained from experimental observations.

For the second problem, it is possible to produce analytical solutions, although very dependent of highly complex computations. Therefore, the interest of the present approach would be mostly of illustration on how a very complex and highly non-linear behaviour can be fairly and efficiently represented by a neural network approach provided that learning data is made available. Furthermore, a replacement of very CPU intensive calculations is enabled, in case of a trained network can be set up.

3.1 Cross sectional behaviour under cyclic loading

The first selected problem is part of an ongoing research effort whose aim is broader than the simple discrimination of sectional properties starting form descriptions of hysteretic behaviour, based on experimental work (Bento and Ndumu, 1996). That overall motivation falls out of the scope of the present text, but may be briefly put forward as the incorporation of trained networks, with "built-in" sectional behaviour, i.e. some kind of "NN-structural black boxes", into programs for non-linear structural analysis (using step by step integration of the dynamic equations of equilibrium).

Such an approach, when successful, enables immense computational gains since the heavy tasks of modelling the sectional behaviour are performed by the trained network, with no explicit modelling of the associated physical phenomena, such as plasticity, fatigue, local buckling and many more.

For simplicity, let us then consider as the problem at hands, the mere identification of the type of section, initiated from the observation of past F-δ charts. In other words, the attempt is to train a network that can recognise the type of cross section – rectangle, circle, I-section or I-composite – based on a description of its hysteretic behaviour fed as input.

A first task when attempting to train any artificial neural network to classify any phenomenon is the identification of previous cases, i.e. of input/output vectors that can be enrolled for training.

In the present case, the available data consisted of experimental outputs produced during

Figure 13 - Column under cyclic loading

sets of cyclic tests of different types of steel and composite cantilever columns such as those of Figure 14 (a). However, since there are many ways in which such diagrams could be passed as input to a NN, three different alternatives were tested: 1) representing the hysteretic diagram using pairs of coordinates of a few (8) characteristic points (Figure 14 b); 2) representing the full contour with a higher and fixed number (50) of equally separated points and 3) using bitmaps containing the full graphical description of the hysteretic behaviour of the sections (Figure 14 c). Additional information about the sections' geometric properties could also be used as input.

For all of the 3 forms of describing the diagrams, 3-layered feedforward networks were used. For all of them the output layer would have a single unit (the one with classification of the section type). In description (1), 16 input nodes (8 x 1 and 8 x 2) were used; in description (2), 100 (50 x 1 and 50 x 2) input nodes were set up; the bitmaps of description (3) originated around 3600 input units, corresponding to a description of the diagram using a bitmap of approximately 5 dpi (dots per inch) in both directions.

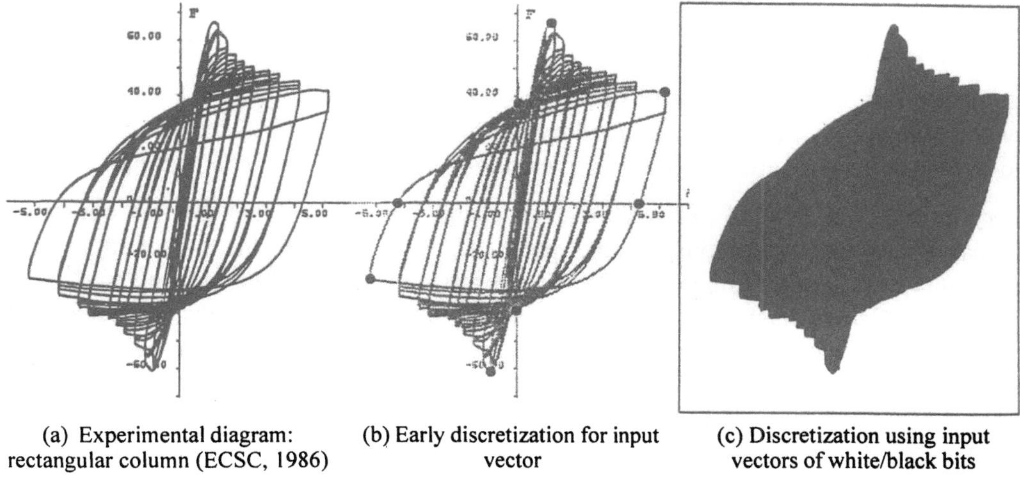

<table>
<tr><td align="center">(a) Experimental diagram:
rectangular column (ECSC, 1986)</td><td align="center">(b) Early discretization for input
vector</td><td align="center">(c) Discretization using input
vectors of white/black bits</td></tr>
</table>

Figure 14 - F *vs.* ∂ hysteretic diagram for rectangular steel column

As for the intermediate layer, various different numbers of units were tested for each of the descriptions. Relatively stable architectures were found at the following topologies:

    (1) 16 x 8 x 1;
    (2) 100 x 3 x 1; (or 106 x 3 x 1, if, for example, 6 geometric properties were added to each input vector as additional sectional information);
    (3) 3528 x 700 x 1.

All three architectures were trained for 3 different types of sections with an extremely low number of training cases – 5 I-sections, 3 rectangular ones and 3 composite ones. 2 I-sections and 1 of each other section were left aside for testing.

The most successful architectures were the first two. Indeed, the 3528 x 700 x 1 did eventually produced a convergent network (one in which the norm of the training error became satisfactorily small) but could not generalise adequately.

One of the most distinctive features of this example is that an indeed very low number of cases was used for training and it was, nevertheless, possible to train several different networks and make them generalise their classification capabilities.

Another interesting feature is that the first descriptions of the diagram – the ones using merely some points (8 or 50) of the diagram's contour – performed much better and immensely faster than the last one, whose bitmap description of the diagram would give rise to an extremely heavier network, thus much harder to train.

The main conclusion regarding this example is that, in general, the results were encouraging, and a methodology for the use of neural networks to address similar applications is worth to develop to a deeper level.

3.2 Frictional contact problem

In this second example, conversely to what happened in the previous one, there exist available analytical solutions (different across the domain), obtained by solving a set of governing equations corresponding to a *conventional mechanical modelling* attitude.

Very briefly, the aim of the problem, as stated here, is to calculate the maximum displacements and or reactions in the point of contact, under the prescription of a horizontal displacement imposed on the contact surface – a mat moving at a given velocity (Figure 15). Those displacements and associated (finite) reactions are the horizontal and vertical ones at the point of contact and the rotation at the connection between the beam and the vertical bar. Since that vertical bar is rigidly connected to a flexible and axially deformable cantilever (with negligible mass), the imposed friction induces a response that is non-linear and requires to be evaluated in the time domain, despite the consideration of geometric linearity. Details of the formulation and analytical solution proposed can be found in Martins and Pinto da Costa (1996).

The analytical resolution of this problem may lead to unstable solutions in which both the evolution of the reactions and displacements at the contact surface, with the prescribed velocity, behave chaotically. Therefore, and given the illustrative nature of this venture, it was decided to produce analytically a small number of solutions, all in a range of stable ones. Moreover, since the integration of the governing solutions requires the use of highly complex symbolic computations, which vary with the actual physical properties of the involved components, it was decided to fix the values of all the physical properties, varying only the prescribed velocity.

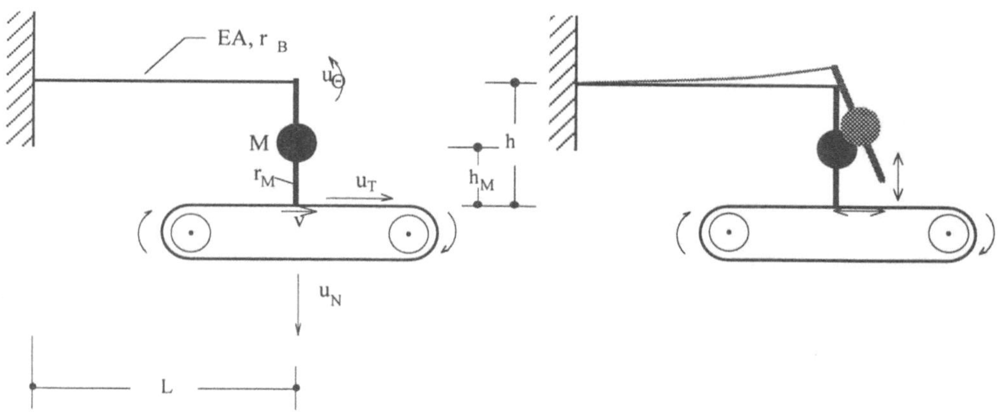

Figure 15 - Pin-on-flat system subject to gravity forces, prescribed displacements and
frictional contact reactions (adapted from Martins and Pinto da Costa, 1996)

Since the main purpose of this example was to illustrate how a trained neural network can
model the non-linear behaviour of a mechanical system (of some complexity) it was decided
to concentrate the modelling activity on the computation of a few representative variables:
the maximum tangent displacement u_T, the maximum normal displacement u_N and the
maximum normal (vertical) reaction r_N.

The specific case for which an analytical solution was produced, corresponds to an
instantiation of the described problem with the following data:

$$\frac{Mg}{EA} = 0.01; \ \frac{R_B}{L} = 0.07; \ \frac{H_B}{L} = 0.1; \ \frac{R_M}{L} = 0.15; \ \frac{H_M}{L} = 0.05; \ \mu=2.2,$$

where M is the mass of the rigid bar, g is the acceleration of gravity, EA is the axial stiffness
of the horizontal beam, R_B and R_M are the ratio of gyration of, respectively, the beam and
the rigid bar cross sections, H_B and H_M are the distances of the mass and the beam to the
contact surface and L is the beam's span and μ is the friction coefficient.

The 3 target quantities were computed for 31 different values of the prescribed velocity.
Therefore, it was possible to produce 31 input/output vectors. Each input vector could
have values for the each of the above quantities, besides that of velocity, with 3 associated
outputs: u_T, u_N and r_N. However, since the only changing input values was that of velocity,
it was decided to set up a network with a single input value.

In order to facilitate the organisation of data and the evaluation of results, the values of
velocity and the output ones were scaled to a fixed interval: for the velocity, the range 1 to
10, while the displacements and reaction were converted to the interval [0.25, 0.75].

Some of the input/output vectors produced analytically – 6 out of 31 – were left aside for
testing of the network, while the remaining 25 ones were all used for training.

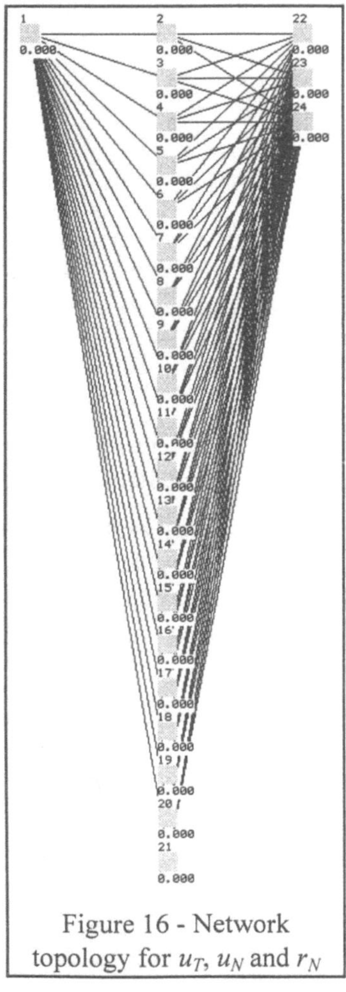

Figure 16 - Network topology for u_T, u_N and r_N

Several networks have been tested, all three-layered with 1 fixed input unit and 1 or 3 output units. Various numbers of hidden units were tested but, generally, for a number of 20 such units, all networks performed well during training. It was decided to train different networks to model separately each of the aimed variables (u_T, u_N and r_N) and also one that could model the 3 quantities at once. Therefore, the actual architectures reported, which synthesise the experiment, are: 1 x 20 x 1, (one for each quantity) and 1 x 20 x 3 (for the 3 variables at once). Figure 16 depicts the network topology for the latter case.

The number of epochs used for each of them for the network to reach an acceptable total is described in Table 2. In this case the error is evaluated as the square root of the square of the errors, SSE, divided by the number of output units. Figure 17 illustrates the progress of the total training and test errors with the number of epochs for the 1x20x3 network referring to the simultaneous modelling of u_T, u_N and r_N.

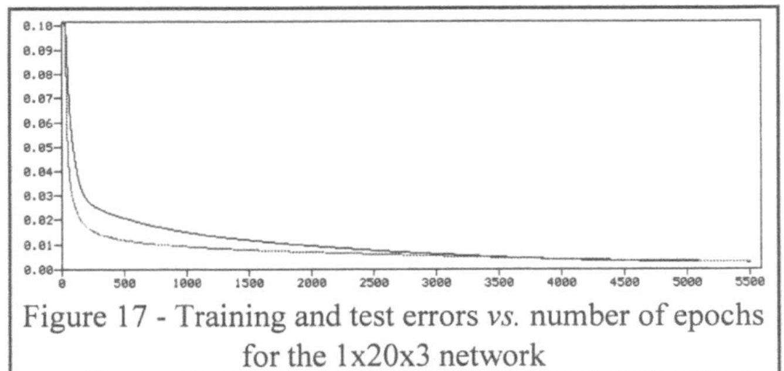

Figure 17 - Training and test errors *vs.* number of epochs for the 1x20x3 network

The same trained networks have been used for estimating their generalisation capacity by computing, through them, the output associated to the test vectors. Table 3 summarises the computed values and the respective errors.

variable	u_T	u_N	r_N	u_T, u_N, r_N
architecture	1 x 20 x 1	1 x 20 x 1	1 x 20 x 1	1 x 20 x 3
no. of epochs	1500	1500	2250	5500
total residual error	0,0029	0,0014	0,0026	0,0021
error per output unit	0,0029	0,0014	0,0026	0,0007

Table 2 - No. of epochs used for training

variable	u_T	u_N	r_N	u_T, u_N, r_N		
			error			
architecture	1 x 20 x 1	1 x 20 x 1	1 x 20 x 1	1 x 20 x 3		
test case 1	0,3%	-0,1%	0,7%	1,3%	1,2%	0,6%
test case 2	-1,3%	-2,0%	-1,2%	-1,2%	-1,4%	-0,5%
test case 3	-0,7%	-0,8%	-1,9%	-2,5%	-2,3%	-1,6%
test case 4	2,4%	3,7%	2,2%	1,4%	1,3%	0,7%
test case 5	4,0%	6,1%	5,8%	6,4%	6,0%	4,5%
test case 6	-2,3%	-2,6%	-0,3%	4,1%	3,3%	3,6%

Table 3 - Errors in test vectors

As seen in Table 3, were the outputs generated by the network for the 6 test cases (cases not used for training), both approaches – the one using one network per variable and the other with a single network modelling the behaviour of the three quantities – have produced very accurate approximations of the analytical one.

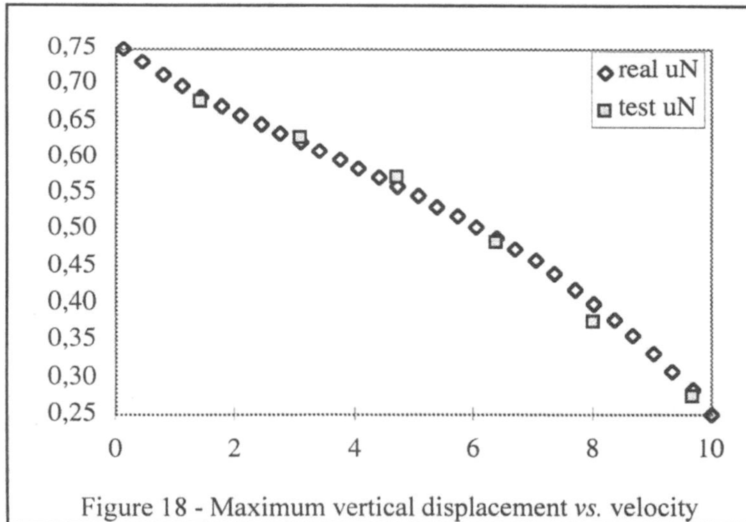

Figure 18 - Maximum vertical displacement *vs.* velocity

Figure 18 to Figure 20 summarise that information, by plotting the analytical prescribed velocity against each of the 3 variables together with the output values computed by the trained neural network. The figure refers to the values computed resorting to the 1x20x3 architectures (the one computing the three values at once).

4. Discussion

The results achieved with the first example enable the prediction that some of the computationally heavy effort associated with step by step non-linear analysis of structures subjected to earthquake loading, namely at the level of the sections' analysis, may eventually be replaced by much faster and equally reliable neural networks trained to replicate the sectional behaviour under cyclic loading (Bento and Ndumu, 1996).

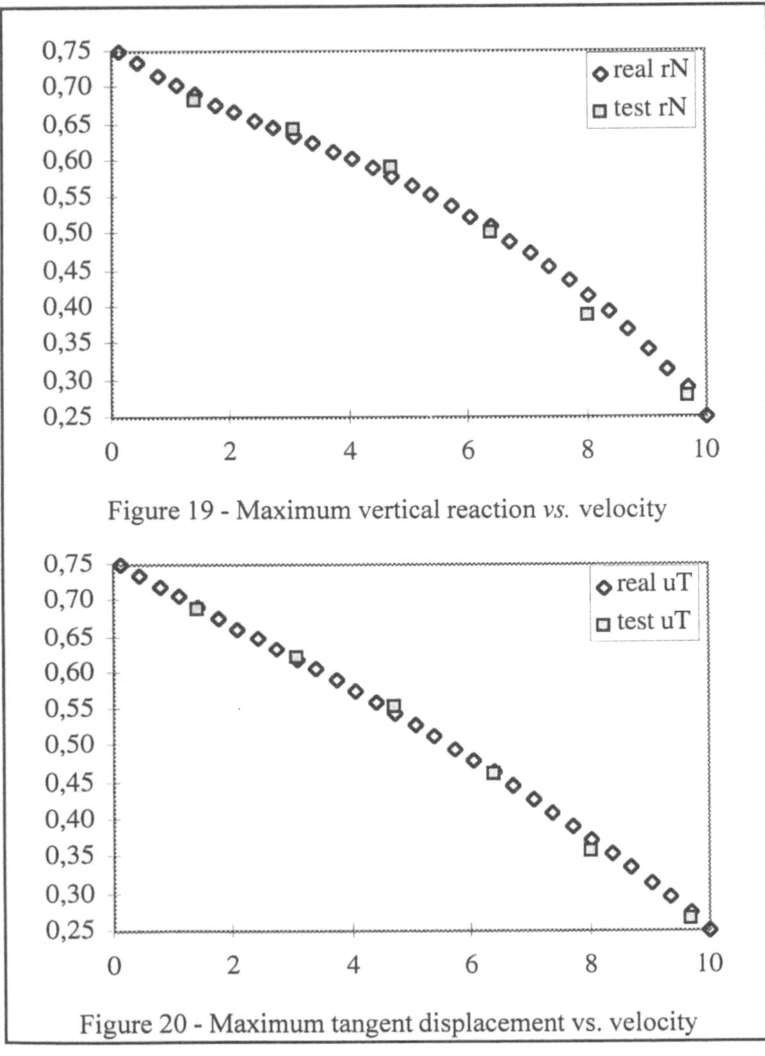

Figure 19 - Maximum vertical reaction *vs.* velocity

Figure 20 - Maximum tangent displacement vs. velocity

For the two examples, it was possible to illustrate that even without any consideration of the physical phenomena involved in a given mechanical system, it is possible to produce very acceptable modelers for those systems, by using artificial neural networks, provided that a history of previous cases, acquired by any means – experimental, analytical, numerical, etc. – may be used for training of the networks.

In the absence of appropriate modelling tools (other than the actual observation of prototypes or physical models) or in the presence of very inefficient ones (for computationally too expensive, for example) the presented approach has demonstrated to be a useful replacement and, sometimes, eventually the only available solution as an effective modelling tool.

Nevertheless, it would be unwise if not ridiculous to face the modelling of mechanical problems with ANN as a generalised replacement for the use of mechanics.

Finally, given the scope and objectives of the present volume, it seems relevant to refer to a few of many references which cover more systematically and in greater depth the application of ANN in structural engineeringand mechanics: Garrett et al. (1993), Flood and Kartam (1994), Takeuchi and Kosugi (1994), Ndumu et al. (1996), Waszczyszyn (1996) provide interesting points of view.

5. Acknowledgements

The present chapter has strongly benefited from a diversified range of influences and inputs from many colleagues of the author. Special thanks are due to Divine Ndumu for his invaluable and continuing contributions since the early phases of exploratory work on the first presented example and to João Martins for his enriching support in the selection of an interesting highly non-linear mechanical problem that would be both complex and didactic. The efforts of António Pinto da Costa in the actual computing of the analytical solutions that produced the input/output vectors used to train the networks of the second example, were especially appreciated.

Thanks are also due to the editors of this book for their careful revision and critique of the text.

6. References

BENTO, J.P., NDUMU, D., "Application of Neural Networks to the Earthquake Resistant Analysis of Structures", *short contribution, EG-SEA-AI 3rd Workshop*, Iain MacLeod (ed.), 111-112, Ross Priory, Scotland, 1996.

ECSC TECHNICAL RESEARCH, *Study on Design of Steel Building in Earthquake Zones*, ECCS, Brussels, Belgium, 1986

FAUSETT, L., *Fundamental of Neural Networks*, Prentice Hall, New Jersey, 1994.

FLOOD, I., KARTAM, N., "Neural Networks in Civil Engineering, I: Principles and understanding; II: Systems and applications", *Journal of Computing in Civil Engineering*, **2(8)**, 131-162, 1994.

GARRETT, J.H., ET AL ., "Engineering Application of Neural Networks", *Journal of Intelligent Manufacturing*, **4**, 1-21, 1993.

HERTZ, J., KROGH, A. and PALMER, R., *Introduction to the theory of neural computation*, Addison-Wesley, 1991.

MARTINS, J.A.C. and PINTO DA COSTA, A., "Stability of finite dimensional systems with unilaeral contact and friction: flat obstacle and linear elastic behaviour", *Report IC-IST AI no.5/96*, Instituto Superior Técnico, 1996.

McCULLOCH, W.S. and PITTS, W., "A Logical Calculus of Ideas Immanent in Nervous Activity", *Bulletin of Mathematical Biophysics*, **5**, 115-133, 1943.

MINSKY, M. and PAPERT, S.A., *Perceptrons*, MIT Press, Cambridge, MA, 1969.

NDUMU, A.N., ET AL. "Simulating Physical Processes with Artificial Neural Networks", International Conference on Engineering Applications of Neural Networks, 9-12, 1996.

ROSENBLATT, F., "The Perceptron: a perceiving and recognizable automaton", *Report 85-460-1*, Project PARA, Cornell Aeronautical Laboratory, Ithaca, New York, 1957.

RUMELHART, D.E., MCLELLAND, J.L and WILLIAMS, R.J., "Learning Internal Representations by Back-Propagating Errors", *Nature*, **323**, 533-536, 1986.

TAKEUCHI, J., Kosugi, Y., "Neural Network Representation of the Finite Element Method", *Neural Networks*, **7(2)**, 389-395, 1994.

Waszczyszyn, Z., "Standard versus refined neural networks applications in civil engineering problems: an overview", *Proceedings of the 2ⁿᵈ Conference on Neural Networks and Their Applications*, 509-516, Czestochowa, Poland, 1996.

WIDROW, B. and HOFF, M.E., "Adaptive switching circuits", in *1960 IRE WESCON Convention Record*, **part 4**, 96-104, New York, 1960.

KNOWLEDGE-BASED SYSTEMS
FOR FINITE ELEMENT MODELING AND INTERPRETATION

S.J. Fenves and G. Turkiyyah
Carnegie Mellon University, Pittsburgh, PA, USA

ABSTRACT

The presentation is largely based on the enclosed report, which appeared in a modified form as a chapter in reprinted from *Research Directions in Computational Mechanics*, U.S. National Committee on Theoretical and Applied Mechanics, National Research Council, (National Academy Press, 1991). The presentation discusses applications in: model generation, model interpretation, integration with design, and comprehensive design environments. Two approaches to interfacing and integrating knowledge-based processes with numerical processes are described and evaluated.

1. AI and Knowledge based systems

Artificial Intelligence (AI) is a branch of Computer Science paralleling other branches which include numerical methods, language theory, programming systems and hardware systems. While computational mechanics has benefited from, and closely interacted with, the latter branches of Computer Science, the interaction between computational mechanics and AI is still in its infancy. Artificial Intelligence encompasses several distinct areas of research each with its own specific interests, research techniques and terminology. These subareas include search technologies, knowledge representation, vision, natural language processing, robotics, machine learning and others.

A host of ideas and techniques from AI can impact the practice of mathematical modeling. In particular, knowledge based systems and environments can provide representations and associated problem solving methods that can be used to encode domain knowledge and domain specific strategies for a variety of ill structured problems in model generation and result interpretation. Advanced AI programming languages and methodologies can provide high level mechanisms for implementing numerical models and solutions, resulting in cleaner, easier to write and more adaptable computational mechanics codes. A variety of algorithms for heuristic search, planning, geometric reasoning, etc. can provide effective and rigorous mechanisms for addressing problems such as shape description and transformation, constraint-based model representation, etc. Before we discuss the applications of AI in mathematical modeling, we briefly review knowledge based and problem solving techniques.

1.1. Knowledge Based Systems

A good standard definition of KBES is the following:

> Knowledge-based expert systems are interactive computer programs incorporating judgment, experience, rules of thumb, intuition, and other expertise to provide knowledgeable advice about a variety of tasks [Gaschnig 81].

The first reaction of many professionals active in computer-aided engineering to the above definition is one of boredom and impatience. After all, conventional computer programs for engineering applications have become increasingly interactive: they have always incorporated expertise in the form of limitations, assumptions, and approximations, as discussed above; and their output has long ago been accepted as advice, not as ''the answer'' to the problem.

There is a need, therefore, to add an operational definition to distinguish the new wave of KBES from conventional algorithmic programs which incorporate substantial amounts of heuristics about a particular application area, or domain. The distinction should not be based on implementation languages or on the absolute separation between domain-dependent knowledge and generic inference engine. The principal distinction lies in the use of knowledge. A traditional algorithmic application is organized into two parts: data and program. An expert system separates the program into an *explicit* knowledge base describing the problem solving knowledge and a control program or inference engine which manipulates the knowledge base. The data portion or context describes the problem being solved and the current state of the solution process. Such an approach is denoted as *knowledge-based* [Nau 83].

Knowledge based systems, as a distinct research area separate from the rest of AI, is about a decade old. This decade of research has seen many changes in the importance placed on various elements of the methodology. The most characteristic change is the methodological one where the focus has shifted from application areas and implementation tools to architectures and unifying principles underlying a variety of problem solving tasks.

In the early days of knowledge based systems, the presentation and analysis of these systems was at two levels. The first level was the level of primitive representation mechanisms (rules, frames, etc.) and primitive inferencing mechanisms associated with them (forward and backward chaining, inheritance and demon firing, etc.) while the second level was the problem description level. Unfortunately, it turned out that the former descriptions are too low-level and do not describe what kind of problem is being solved while the latter descriptions are necessarily domain specific and often incomprehensible and uninteresting for people outside the specific area of expertise.

What is needed then is a description level that can adequately describe what heuristic programs do and know — a computational characterization of their competence independent of task domain and independent of programming language implementation. Several characterizations of generic tasks that arise in a multitude of domains have been presented in [Clancey 85, McDermott 88, Chandrasekaran 86]. Generic tasks are described by the kind of knowledge they rely on and their control of problem solving. Generic tasks constitute higher level building blocks for expert systems design and their characterizations form the basis for analyzing the contents of a knowledge base (for completeness, consistency, etc.), for describing the operation and limitations of systems and for building specialized knowledge acquisition tools.

1.2. Problem Solving

Many problem solving tasks can be formulated as a search in a state space. A state space consists of all the states of the domain and a set of operators that change one state into another. The states can best be thought of as nodes in a connected graph and the operators as edges. Certain nodes are designated as goal nodes and a problem is said to be solved when a path from an initial state to a goal state has been found. State spaces can get very large and various search methods to control the search efficiency are appropriate.

- Search reduction. This technique involves showing that the answer to a problem cannot depend on searching a certain node. There are several reasons this could be true: (1) No solution can be in the subtree of this node. This technique has been called "constraint satisfaction" and involves noting that the conditions that can be attained in the subtree below a node are insufficient to produce some minimum requirement for a solution; (2) Solution in another path is superior to any possible solution in the subtree below this node; and (3) Node has already been examined elsewhere in the search. This is the familiar dynamic programming technique in operations research.

- Problem reduction. This technique involves transforming the problem space to make searching easier. Examples of problem reduction include: (1) Planning with macro operators in an abstract space before getting down to the details of actual operators; (2) means-end analysis which attempts to reason backward from a known goal; and (3) sub-goaling which decomposes difficult goals into simpler ones until easily solved ones are reached.

- Adaptive search techniques. These techniques use evaluation functions to expand the "next best" node. Some algorithms (A^*) expand the node most likely to contain the optimal solution. Others (B^*) expand the node that is most likely to contribute the most information to the solution process.

- Using domain knowledge. One way of controlling the search is to add additional information to non-goal nodes. This information could take the form of a distance from a hypothetical goal, operators that may be usefully applied to it, possible backtracking locations, similarity to other nodes which could be used to prune the search or some general goodness information.

2. Applications in Mathematical Modeling

Mathematical modeling is the activity devoted to the study of the simulation of physical phenomena by computational processes. The goal of the simulations is to predict the behavior of some artifact to its environment. Mathematical modeling subsumes a number of activities as illustrated by Figure 1 below.

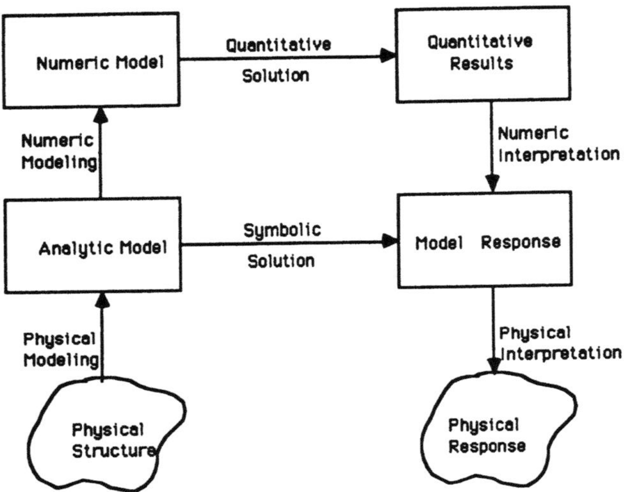

Figure 2-1: Mathematical Modeling Process

The following sections discuss the applications and potential impacts of AI technology on the various mathematical modeling activities. The mathematical modeling activities presented include model generation, interpretation of numerical results and development and control of numerical algorithms. It is to be noted that these activities are not independent and this organization is used primarily to assist in the exposition of ideas.

2.1. Model Generation

We use the term model generation to encompass all activities that result in the generation of models of physical systems suitable as input for a computational mechanics program. The generation of mathematical models from physical descriptions of systems is a problem of great practical importance. In all disciplines that use computational mechanics —aerospace, nuclear, marine, civil and mechanical— there is a need for modeling an increasingly wider range of phenomena in all stages of system design, from the earliest conceptual studies to the most detailed component performance evaluation. In addition, there is an urgent need for much closer integration of computational mechanics evaluations into computer aided design and in extending analyses to computer aided manufacturing where there is great interest in analyzing not just finished components, structures or systems, but the manufacturing processes themselves, such as casting, forging or extrusion.

With the availability of literally hundreds of computational mechanics codes including a large number of general-purpose finite element programs with a broad range of capabilities, model generation has become the primary activity of the analyst. However, in the current state of the art, the preparation of input data is a tedious, error prone and time consuming process. Analysts are forced to interact with programs at a level much lower than the conceptual level at which they make decisions. Hence, there is a need for higher level interfaces to programs that can free analysts from details, allow them to generate models in terms of high level behavioral descriptions and thereby increase their productivity and improve the quality and reliability of their analyses.

Moreover, because of the very small number of experienced modelers who can confidently and reliably model physical problems and the increasing need for modeling expertise, it has also become increasingly important to capture and organize the modeling knowledge of experienced analysts and transfer it to less experienced and novice analysts. Some of the dangers that will ensue if this transfer of knowledge and attitude do not occur have been cogently argued in Smith [Smith 86]. The methodology of artificial intelligence and knowledge based systems promises to provide an opportunity to respond to the needs identified above.

We discuss model generation tools at three levels of increasing abstraction.

2.1.1. Intelligent Help Systems

Intelligent help systems address the issue of providing consulting advice to non expert engineers. The subject of help could either be how to use a particular analysis program or what model parameters and procedures are appropriate for particular physical systems [Bennett 78, Gaschnig 81, Taig 86a, Cagan 87]. Help systems are not connected to analysis programs and are not meant to provide complete solutions to modeling problems. They simply guide the user — typically the novice user — in conducting some modeling tasks.

Help systems typically act as interactive passive consultants. They query the user on some key aspects of the problem, and based on the key problem features inform the user on the appropriate sequence of commands to use, program options to select, analysis strategies to invoke, numerical parameters to assign, etc. The interaction is often through a question and answer session and custom menus. These help systems can be readily built using simple shells that provide forward and/or backward chaining capabilities. With the advent of powerful PC software for writing and organizing knowledge and communicating with the user through standard interfaces (HyperCard, HyperX) such systems can be properly integrated in a variety of analysis and design environments.

2.1.2. Customized Preprocessors

Customized preprocessors are knowledge based programs that are integrated into the environment they operate in [Zumsteg 85, Gregory 86, Reynier 86]. Customized preprocessors extract relevant features from a data base describing the physical object to be modeled (often a simple geometric model). These features play the role of higher level, symbolic descriptions that provide semantics to geometric entities. Features are used to classify various components and match them to corresponding analysis methods and parameters (e.g., finite element mesh density). These parameters are then used to drive special purpose interfaces to produce input files for the appropriate analysis programs of the environment.

The advantage of customized systems is that the user intervention in the modeling process is minimal. Essentially, the user is only required to enter some description of the physical object to be modeled. The preprocessors rely on the fact that the object to be modeled can be adequately described in terms of a predetermined set of features encoded in the knowledge base and that a set of rules for modeling, analyzing and evaluating these features exist. Clancey [Clancey 85] has analyzed the structure of this class knowledge based systems in terms of three primitive tasks: (1) data abstraction (definitional, qualitative or generalization abstractions); (2) heuristic associations between the abstracted features that characterize the object and the categories of known models; and (3) refinement that specializes a chosen model category to the particular problem at hand.

Unfortunately, customized preprocessors are typically limited to narrow domains. This is due to the fact that they rely on structuring the objects of the domain in very specific ways: to fit the templates of a set of a-priori chosen features. The ways in which models can be used must be anticipated and fixed at system-design time. As the domain expands, significant knowledge engineering effort is required to find, organize and encode the myriad of little pieces of knowledge needed to extract all relevant features and analyze their potential interactions. The combinatorial explosion of rules needed to cover very large domains can become prohibitive.

2.1.3. High Level Generation Tools

High level model generation tools incorporate techniques that are more flexible than the heuristic classification approaches used by the systems discussed above. In particular, the goal of these tools is to put at the disposal of the analyst a set of powerful representations and reasoning mechanisms that can be invoked as needed and that serve as means of high level specifications of modeling operations.

A representation of modeling knowledge that can provide effective modeling assistance is an explicit representation of the assumptions that can generate various model fragments. Assumptions are the basic entities from which models are constructed, evaluated and judged adequate. Analysts often refer to and distinguish various models by the assumptions incorporated in them. The vocabulary of assumptions corresponds more closely to how analysts describe and assess the limitations of proposed models. Hence, the explicit representation and use of modeling assumptions in a modeling assistant can make the modeling operations correspond more closely to analysts' methods and could make it easier to organize and build a knowledge base. Assumptions encode a larger chunk of knowledge than rules and hence can provide a conceptual structure that is clearer and easier to manage ·than the typical knowledge base of rule-based systems.

An example of how assumptions can be represented in a modeling assistant is provided in [Turkiyyah 90]. In this representation, assumptions are modular units that incorporate, besides a prescription of how they affect the model, declarative knowledge about the scope of their applicability, their relevance for various modeling contexts as well as their (heuristic) a-priori and (definitional) a-posteriori validity conditions. Assumptions can be either used directly by the analyst or indirectly through analysis objectives. When an analysis objective is posted, a planning algorithm selects an appropriate set of assumptions that can satisfy the modeling objective. These assumptions can then be automatically applied to generate a model that can be input to a finite element program.

Another example of a high level model generation tool is discussed in [Turkiyyah 90]. The central idea is to generate from a geometric description of an object an abstraction —the skeleton— that can capture intrinsic object characteristics. The skeleton is effectively a symbolic representation of shape that corresponds to how analysts seem to visualize and describe shape and shape information, namely in terms of axis and width for elongated subdomains, center and radii for rounded subdomains, bisector and angle for pointed subdomains, etc. Because such abstractions are domain-independent and hence general purpose, they can be used to suggest simplifications to the model (e.g., replace certain elongated two-dimensional regions by one-dimensional beam elements). They can also be used to subdivide a spatial domain into subregions of simple structure that can then be meshed directly by mapping or other mesh generation techniques.

2.2. Model Interpretation

This section describes the potentials of AI for assisting in post-analysis operations. Post-analysis operations are generically referred to as *interpretation*, although they involve distinctly different types of processes including model validation, response abstraction, response evaluation and redesign suggestion. We examine these areas next.

2.2.1. Model Validation

Model validation the task of assessing whether the numerical results of the mathematical model can be confidently believed to reproduce the real behavior of the system being modeled. Knowledge based techniques provide practical mechanisms to represent and characterize one important class of possible errors — idealization errors. Turkiyyah and Fenves [Turkiyyah 90] have proposed a framework for validation of idealized models. The main idea is that if the model of a system is systematically generated through the application of a set of assumptions on the system's representation, then any idealization error can be traced to one or more of those generative assumptions. Furthermore, each assumption encodes the conditions under which it is valid, hence model validation involves checking the validity conditions of individual assumptions. There are many ways of verifying assumptions ranging in complexity from the evaluation of simple algebraic expressions to the analysis of a model that is more detailed than the original one.

2.2.2. Abstraction of Numerical Results

Response abstraction is the task of generating some abstract description of the raw numerical results obtained from the analysis program. This description is presented in terms of high-level aggregate quantities representing key response parameters or behavior patterns.

Response abstractions can be classified in two types. The first type, functional response abstractions, depends on the role that various subsystems or components play outside the model proper, i.e., the meaningful aggregate quantities that are generated depend on knowledge of characteristics beyond the geometric and material properties of the system. The ability to generate the function-dependent response abstraction depends on the ability to represent the functional information that underlies the object being modeled.

The second type of response abstraction is function-independent. One seeks to recognize patters, regularities and interesting aspects, and generate qualitative descriptions (e.g., stress paths) of the numerical results, independent of the functional nature of the object being modeled. Techniques from computer vision and range data analysis can be used to generate these interpretations. Well developed vision techniques such as aggregation, segmentation, clustering, classification, recognition can be applied to the task. One interesting use of response abstractions is in assisting the user in checking the "physical reasonableness" of the numerical results, by comparing response abstractions from more refined models to the response of simpler models.

2.2.3. Conformance Evaluation

Conformance evaluation is the task of verifying that the computed results satisfy design specifications and functional criteria such as stress levels, ductility requirements or deflection limitations. Conformance evaluation is largely a diagnostic problem and the well developed techniques for diagnosis can be applied to the task. Conformance evaluation requires heuristics on: (1) what are possible failure modes; (2) what are the applicable code and standard provisions; (3) what response quantities (stresses, deflections, etc) are affected by the provisions; and (4) what approximations of the provisions and responses are necessary.

A major issue in developing expert systems for conformance evaluation is that of representing code and standard provisions in a form suitable for evaluation, yet amenable to modification when standards change or when an organization wishes to use its own interpretations. One technique suitable for this purpose is to represent standards provisions by networks of decision tables [Fenves 79]. The use of this representation in an expert system environment is demonstrated in [Garrett 89].

2.2.4. Integration in Design

Analysis is rarely, if ever, an end in itself: the overwhelming use of analysis is to guide and confirm design decisions. One important application of AI techniques is in providing redesign recommendations when the response of the system analyzed is not satisfactory. One problem that has to be addressed concerns the nature of the knowledge base that can generate redesign recommendations: should it be separate and independent, or should it use the same modeling knowledge responsible for generating and interpreting models. In a *generative* modeling system, the second approach formulates redesign as the following goal-oriented problem: given some deficiencies uncovered by an analysis, what modifications to the design object are required so that a model whose response satisfies the design specifications can be generated.

Another problem that has to be addressed, if computational mechanics methods are to be be adequately incorporated in Computer Aided Design, is a general capability for providing analysis interpretations and design evaluations compatible with the progress of the design process from an initial conceptual sketch to a fully detailed description. Evaluations should occur at increasingly higher degrees of refinements throughout the design process. Initial simple models can provide early feedback to the designer (or design system) on the global adequacy of the design, while evolved models, paralleling the evolving design, help to guide the designer in the detailed design stages. An important issue in developing general mechanisms for hierarchical modeling is how to generate and represent various kinds of geometric abstractions.

2.3. Numerical Model Formulation

We use the term formulation to denote the process of producing a computational mechanics capability —a set of numerical routines— from a representation of some physical phenomenon. It is well known that the development of a computational mechanics program is time consuming, expensive and error prone. Processes that can help in the quick development of reliable numerical software can be of great practical benefit. Ideas from AI can significantly contribute to various aspects of the formulation process: performing symbolic computations, expressing subroutines in a form that make them reusable, designing large systems with appropriate data abstractions, assisting in the synthesis of computational mechanics programs and integrating heuristics and knowledge based methods into numerical solutions. We examine these potentials in turn.

2.3.1. Symbolic Processing

One aspect of program development that is particularly time consuming and error-prone is the transition from a continuum model, involving operations of differentiation and integration, to a computational model, involving algebraic and matrix operations. A branch of AI deals with symbolic computations, culminating in symbolic computation programs such as MACSYMA and Mathematica. Programs in this class operate on symbolic expressions, including operations of differentiation and integration, producing resulting expressions in symbolic form. A particularly attractive practical feature of these programs is that the output expressions can be displayed in FORTRAN source code format.

The potential role of symbolic processing has been investigated by several researchers. Particularly thorough surveys and evaluation papers appear in [Noor 79] and [Noor 81]. Recent work by Wang and his colleagues concentrates on efficient FORTRAN source code generation [Wang 85]. These studies indicate that symbolic processing can significantly assist in the generation of computational model components to be incorporated in the source code. Symbolic generation of source code for element stiffness and load matrices can eliminate the tedious and error-prone operations involved in going from a differential, algebraic representation to a discrete, procedural representation. Additional run-time efficiency improvements are possible through functionally optimized code and the use of exact integrations. Finally, conceptual improvements are possible, such as symbolically condensing out energy terms that contribute to shear and membrane locking.

2.3.2. Reusable Subroutines

Numerical subroutines that perform function evaluations, domain and boundary integrations, linear and non-linear equation solving, etc. abound in computational mechanics codes. However, the typical implementation of these subroutines bear little resemblance to our mathematical knowledge of the operations that these subroutines perform. They are written as a sequence of concrete arithmetic operations that include many mysterious numerical constants and are tailored to specific machines. Because these routines do not exhibit the structure of the ideas from which they are formed, their structure is monolithic, hand crafted for the particular application —rather than constructed from a set of interchangeable parts that represent the decompositions corresponding to the elemental concepts that underly the routine. Such numerical routines are often difficult to write and even more difficult to read.

Roylance [Roylance 88] shows examples of how even the simplest routines that are often thought of as "atomic" —such as sin(x)— can be constructed from their primitive constituent mathematical operations, i.e., periodicity and symmetry of the sine function, and a truncated Taylor expansion. Abelson [Abelson 89] shows how Romberg's quadrature can be built by combining a primitive trapezoidal integrator with an accelerator that speeds the convergence of a sequence by Richardson extrapolation. The idea is that instead of writing a subroutine that computes the value of a function, one writes code to construct the subroutine that computes a value.

Such a formulation separates out the ideas into several independent pieces that can be used interchangeably to facilitate attacking new problems. The advantages are obvious. First, clever ideas need be coded once in a context independent of the particular application, thus enhancing the reliability of the software. Second, the code is closer to the mathematical basis of the function and is expressed in terms of the vocabulary of numerical analysis. Third, the code is adaptable to various usages and precisions because the routine's accuracy is an integral part of the code rather than a comment that the programmer adds: just changing the number that specifies the accuracy will generate the single, double and quadruple precision versions of a subroutine.

Writing subroutines in this style requires the support of a programming language that provides higher order procedures, streams and other such powerful abstraction mechanisms, as available in functional languages. Roylance shows that the run time efficiency does not suffer. The extra work of manipulating the construction of the function need be done only once. The actual calls of the function are not encumbered. Moreover because functional programs have no side effects they have no required order of execution. This makes it exceptionally easy to execute them in parallel.

2.3.3. Programming with Data Abstractions

The current generation of computational mechanics software is based on programming concepts and languages two or three decades old. As attention turns to the development of the next generation software, it is important that the new tools, concepts and languages that have emerged in the interim be properly evaluated and that the software be built using the best of the appropriate tools.

Baugh and Rehak [Baugh 89] presented a design for a finite element system based on object oriented concepts. They showed how object-oriented programming, an offshoot of AI research, can have a major impact on computational mechanics software development. In particular, they showed how it is possible to raise the level of abstraction present in large-scale scientific programs (i.e., allowing finite element programmers to deal directly with concepts such as elements and nodes) by identifying and separating levels of concern. Programs designed in this manner allow developers to reason about program fragments in terms of abstract behavior instead of implementation. These program fragments are referred to as objects or data abstractions, their abstract quality being derived from precise specifications of their behavior that is separate and independent of implementation and internal representation details.

2.3.4. Model Synthesis Assistance

While the bulk of today's computational mechanics production work is done by means of large, comprehensive programs, there is a great deal of exploratory work requiring the development of "one-shot" ad-hoc custom-built programs. Developers of such ad-hoc programs may have access to subroutine libraries for common modules or "building blocks", but not much else. These developers frequently have to re-implement major segments of complete programs so as to be able to "exercise" the few custom components of their intended program.

One potential application of AI methodology is an expert system to assist in synthesizing computational programs, tailored to particular problems, on the fly. The system would require as input some specifications of the goal of the program, the constraints (e.g., language, hardware environment, performance constraints, etc.) and the description of the custom components (e.g., a new equation solver, a new element, a new constitutive equation for a standard element). The system's knowledge base would contain descriptions of program components with their attributes (language, environment, limitations, interface descriptions, etc.) and knowledge about combining program components, including possibly knowledge for writing interface programs between incompatible program segments. The expert system would have to use both backward chaining components and forward chaining components: the former to decompose the goal into the program structure, the latter for selecting program components to "drive" the low-level custom components. A prototype system for synthesizing and customizing finite element programs has been presented by Nakai [Nakai 89].

2.3.5. Monitoring Numerical Solutions

Combining numerical techniques with ideas from symbolic computation and with methods incorporating knowledge of the underlying physical phenomena can lead to a new category of intelligent computational tools for use in analysis. Systems that have knowledge of the numerical processes embedded within them and can reason about the application of these processes, can control the invocation and evolution of numerical solutions. They can "see what not to compute" [Abelson 89] and take advantage of known characteristics of the problem and structure of the solution to suggest data representations and appropriate solution algorithms.

The coupling of symbolic (knowledge based) and numerical computing has been the subject of two recent workshops [Kowalik 86, Kowalik 87]. The primary motivation for coupled systems is to support situations where the application of pure numerical approaches does not provide the capabilities needed for a particular application. Frayman [Frayman 87] couples numerical function minimization methods with constraint based reasoning methods from AI technology to successfully attack a problem space that is large, highly non-linear and where numerical optimization methods are too weak to find a global minimum. This problem is typical of many large interdependent constraint satisfaction problems found in engineering models. Domain specific knowledge about problem solving in terms of symbolic constraints guide the application of techniques such as problem decomposition, constraint propagation, relaxation and refinement to derive a solution to the problem.

2.4. Comprehensive Modeling Environments

As higher level modeling tools are built and larger modeling knowledge bases constructed, the issues of integration, coordination, cooperative development, customization, etc. become critical. Fenves [Fenves 85, Fenves 86] has suggested a framework for a general finite element modeling assistant. The framework is intended to permit a cooperative development effort involving many organizations. The key feature of the framework is that the system consists of a set of *core* knowledge sources for the various aspects of modeling and model interpretation which use stored *resources* for the problem dependent aspects of the task. In this fashion, new problem types, as well as individual organizations' approaches approaches to modeling, involve only expansion of the resources without affecting the knowledge sources.

In the comprehensive framework envisaged, the core knowledge sources would perform the functions of model generation and model interpretation discussed in Sections 2.1 and 2.2, and the function of program selection (with possible customization and synthesis as discussed in Section 2.3.4) and invocation. The three major resources used by these knowledge sources are as follows:

- Physical class taxonomies. These represent an extended taxonomy or semantic network of the various classes of physical systems amenable to finite element modeling and the assumptions appropriate for each class. Their purpose is to provide pattern matching capabilities to the knowledge sources so that the definition of problem class and key problem parameters can be used by the knowledge sources in their tasks at each level of abstraction. The major design objective in developing these taxonomies will be to avoid exhaustive enumeration of individual problems to be encountered, but rather to build a multi-level classification of problem types based on their functionality, applicable assumptions, behavior, failure modes, analysis strategies and spatial decompositions. It is also expected that a large part of knowledge acquisition can be isolated into modifying these taxonomies either by specialization (customization to individual organization) or generalization (merging or pooling knowledge of separate organizations).

- Program capability taxonomies. These represent, in a manner similar to the above, the capabilities, advantages and limitations of analysis programs. The taxonomy must be rich enough so that the knowledge source that invokes the programs can make recommendations on the appropriate program(s) to use based on the high level abstractions generated by the other knowledge sources, or, if a particular program is not available in the integrated system, make recommendations on alternate modeling strategies so that the available program(s) can be effectively and efficiently used. As the previous taxonomy, the program capability taxonomy needs to be designed so that knowledge acquisition about additional programs can be largely isolated to the expansion of the taxonomy data base.

- Analysis programs. The programs, including translators to and from neutral files as needed, are isolated in the design to serve only as resources to solve the model. The issues in this interconnection are largely ones of implementation in coupling numerical and knowledge based programs. Modern computing environments make such coupling relatively seamless.

3. Research Issues

This section attempts to briefly discuss two important problems that have to be addressed before reliable modeling environments such as the one discussed above can be built. The first problem is the need for providing more flexibility to knowledge based systems and the second is the need for compiling a core of modeling assumptions.

3.1. Flexible Knowledge Based Systems

The present generation of knowledge based systems has been justly criticized on three grounds: that they are brittle, idiosyncratic, and static.

Present knowledge based systems are brittle — in the sense used in computer science as a contrast to "rugged" systems — in that they work in a very limited domain and fail to recognize, much less solve, problems falling outside of their knowledge base. In other words, these systems do not have an explicit representation of the boundaries of their expertise. Therefore, there is no way for these systems to recognize a problem for which their knowledge base is insufficient or inappropriate. Rather than exhibiting "common sense reasoning" or "graceful degradation", the systems will blindly attempt to "solve" the problem with their current knowledge, producing predictably erroneous results. Current research on reasoning from first principles will help overcome this problem. Combining first principles with specialized rules will allow a system to resort to sound reasoning when few or no specialized items in its knowledge base cover a situation. First principles can also be used to check the plausibility of conclusions reached by using specialized knowledge.

A KBES developed using the present methodology is idiosyncratic in the sense that its knowledge base represents the expertise of a single human domain expert or, at best that of a small group of domain experts. The system thus reproduces only the heuristics, assumptions, and even style of problem solving of the expert or experts consulted. It is the nature of expertise and heuristics that another, equally competent expert in the domain may have different, or even conflicting, expertise. However, it is worth pointing out that a KBES is useful to an organization only if it reliably reproduces the expertise of that organization. At present, there appear to be no usable formal methods for resolving the idiosyncratic nature of KBES. There are some techniques for checking the consistency of knowledge bases, but these techniques are largely syntactic. One practical approach is to build a domain-specific *meta-shell* which contains a common knowledge base of the domain and excellent knowledge acquisition facilities for expansion and customization by a wide range of practitioners.

Present KBES are static in two senses. First the KBES reasons on the basis of the current contents of its knowledge base; a separate component the knowledge acquisition facility is used to add to or modify the knowledge base. Second, at the end of the consultation session with a KBES, the context is cleared, so that there is no provision for retaining the "memory" of the session (e.g., the assumptions and recommendations made). Research on machine learning is maturing to the point where knowledge based systems will be able to learn by analyzing their failed or successful performance — an approach sometimes called explanation based learning [Minton 89]. Learning by induction from a large library of solved cases can also allow systems to learn classification rules [Reich 89].

3.2. Compilation of Modeling Knowledge

One task of great practical payoff is the development of a knowledge base of modeling assumptions, that contains what is believed to be the shared knowledge of analysts. Such a core knowledge base will be beneficial in two important ways. First, it could be used as a starting point to build a variety of related expert systems, hence making the development cycle shorter. Second, such a knowledge base could become the "corporate memory" of the discipline and hence could give us insights into the nature of various aspects of modeling knowledge. One starting point to build such a knowledge base is to "reverse engineer" existing models to recognize and extract their assumptions.

Two useful precedents from other domains offer guidance. Cyc [Lenat 88] is a large scale knowledge base intended to encode knowledge spanning human consensus reality down to some reasonable level of depth — knowledge that is assumed to be shared between people communicating in everyday situations. Cyc is a 10-year effort that started in 1984 and progress to date indicate that the already very-large KB (millions of assertions) is not diverging in its semantics and already can operate in some common situations.

KBEmacs [Waters 85], for Knowledge-Based Emacs, is a programmer's apprentice. KEmacs extends the well known text editor Emacs with facilities to interactively support programming activities. The knowledge base of KBEmacs consists of a number of abstract program fragments — called *cliches* — ranging from very simple abstract data types such as lists to abstract notions such as synchronization and complex subsystems such as peripheral device drivers. The fundamental idea is that the knowledge base of cliches encode the knowledge that is believed to be shared by programmers. KBEmacs has been used successfully to build medium size programs in Lisp and Ada.

4. Closure

The objective of this paper was to present some of the concepts and methodologies of artificial intelligence and examine some of their potential applications in various aspects of mathematical modeling. The methodologies sketched in this paper are maturing rapidly and many new applications in mathematical modeling are likely to be found. Undoubtedly, AI methodologies will eventually become a natural and integral component of the set of computer based tools of engineers, to the same extent as present day "traditional" algorithmic tools. These tools will then significantly elevate the role of computers in engineering from the present-day emphasis on *calculation* to the much broader area of *reasoning*.

References

[Abelson 89]   H. Abelson et al., "Intelligence in Scientific Computing," *Communications of the ACM*, Vol. 32, No. 5, pp. 546-562, May 1989.

[Baugh 89]   J. W. Baugh and D. R. Rehak, *Computational Abstractions for Finite Element Programming*, Technical Report R-89-182, Carnegie Mellon University, Pittsburgh, PA 15213, September 1989.

[Bennett 78]   J. Bennett et. al., *SACON: A Knowledge-based Consultant For Structural Analysis*, Technical Report STAN-CS-78-699, Stanford Heuristic Programming Project, 1978.

[Cagan 87]   J. Cagan and V. Genberg, "PLASHTRAN: An Expert Consultant on Two-dimensional Finite Element Modeling Techniques," *Engineering with Computers*, 1987.

[Chandrasekaran 86]
                 B Chandrasekaran, "Generic Tasks in Knowledge Based Reasoning," *IEEE Expert*, Fall 1986.

[Clancey 85]   W. J. Clancey, "Heuristic Classification," *Artificial Intelligence*, Vol. 27, No. 3, 1985.

[Fenves 79]   S. J. Fenves, "Recent Developments in the Methodology for the Formulation and Organization of Design Specifications," *Engineering Structures*, Vol. 1, pp. 223-229, October 1979.

[Fenves 85]   S. J. Fenves, "A Framework for a Knowledge-Based Finite Element Assistant," *Applications of Knowledge Based Systems to Engineering Analysis and Design*, Clive L. Dym, Ed., ASME, 1985.

[Fenves 86]   S. J. Fenves, "A Framework for Cooperative Development of a Finite Element Modeling Assistant," *Reliability of Methods for Engineering Analysis*, K. J. Bathe and D. R. J. Owen, Ed., Pineridge Press, Swansea, U.K, pp. 475-486, 1986.

[Frayman 87]   F. Frayman, "Solving Large Scale Interdependent Constraint Satisfaction Problems," *Coupling Symbolic and Numeric Computing in Expert Systems*, J. S. Kowalik and C. T. Kitzmiller, Ed., North-Holland, 1987.

[Garrett 89]   J. H. Garrett and S. J. Fenves, "Knowledge Based Standard-Independent Member Design," *Journal of Structural Engineering*, Vol. 115, No. 6, June 1989.

[Gaschnig 81]   J. Gaschnig, R. Reboh and J. Reiter, *Development of a Knowledge Based System for Water Resource Problems*, Technical Report 1619, SRI International, August 1981.

[Gregory 86]   B. L. Gregory and M. S. Shephard, "Design of a Knowledge Based System to Convert Airframe Geometric Models to Structural Models," *Expert Systems in Civil Engineering*, C. N. Kostem and M. L. Maher, Ed., ASCE, 1986.

[Halfant 88]   M. Halfant and G. J. Sussman, "Abstraction in Numerical Methods," *Proceeding of ACM Conference on Lisp and Functional Programming*, Aug, 1988.

[Kowalik 86]   J. S. Kowalik, Ed., *Coupling Symbolic and Numerical Computing in Expert Systems*, North-Holland, 1986.

[Kowalik 87]   J. S. Kowalik and C. T. Kitzmiller, Ed., *Coupling Symbolic and Numerical Computing in Expert Systems, II*, North-Holland, 1987.

[Lenat 88]   D. Lenat and H. Guha, *The World According to CYC*, Technical Report ACA-AI-300-88, MCC, September 1988.

[McDermott 88]   J. McDermott, "Towards a Taxonomy of Problem Solving Methods," *Automating Knowledge Acquisition for Expert Systems*, S. Marcus, Ed., Klwer Academic Publishers, 1988.

[Meyer 84]   Christian Meyer and John McCormick, "Mathematical Modeling Of Complex Structures For Dynamic Analysis," *Computers and Structures*, Vol. 18, No. 4, pp. 673-688, 1984.

[Meyer 87]   C. Meyer, editor, *Finite Element Idealization*, ASCE, 1987.

[Minton 89]   S. Minton et al., *Explanation-Based Learning: A Problem Solving Perspective*, Technical Report CMU-CS-89-103, Carnegie Mellon University, January 1989.

[Nakai 89]   S. Nakai, *A Knowledge Based Approach to Engineering Program Synthesis*, Technical Report R-89-183, Civil Engineering Department, Carnegie Mellon University, October 1989.

[Nau 83]   D. S. Nau, "Expert Computer Systems," *IEEE Computer*, Vol. 16, February 1983.

[Noor 79]   A. K. Noor and C. M. Anderson, "Computerized Symbolic Manipulation in Structural Mechanics: Progress and Potential," *Computers and Structures*, Vol. 10, 1979.

[Noor 81]   A. K. Noor and C. M. Anderson, "Computerized Symbolic Manipulation in Non-Linear Finite Element Analysis," *Computers and Structures*, Vol. 13, 1981.

[Reich 89]   Y. Reich and S. J. Fenves, "The Potential of Machine Learning Techniques for Expert Systems," *(AI EDAM)*, Vol. 3, No. 3, 1989.

[Reynier 86]   M. Reynier, "Interactions between Structural Analysis, Know-How and Chain of Reasoning used by the CARTER Expert System for Dimensioning," *Reliability of Methods for Engineering Analysis*, K. J. Bathe and D. R. J. Owen, Ed., Pineridge Press, Swansea, U.K, 1986.

[Roylance 88]   G. Roylance, "Expressing Mathematical Subroutines Constructively," *Proceeding of ACM Conference on Lisp and Functional Programming*, pp. 8-13, Aug, 1988.

[Smith 86]   G. Smith, "The dangers of CAD," *Mechanical Engineering*, Vol. 108, No. T63-2, February 1986.

[Taig 86]   I. C. Taig, "Expert Aids to Finite Element System Applications," *Applications of Artificial Intelligence to Engineering Problems*, D. Sriram and R. Adey, Ed., Springer-Verlag, 1986.

[Turkiyyah 90]   G. Turkiyyah and S. J. Fenves, *Generation and Interpretation of Finite Element Models in a Knowledge Based Environment*, Technical Report to appear, Civil Engineering Department, Carnegie Mellon University, 1990.

[Wang 85]   P. S. Wang, "Automatic Derivation and Generation of Fortran Programs for Finite Element Analysis," *Proceedings ISA*, 1985.

[Waters 85]   R. C. Waters, "The Programmer's Apprentice: A session with KBEmacs," *IEEE Transactions on Software Engineering*, Vol. 11, No. 11, 1985.

[Zumsteg 85]   J. R. Zumsteg and D. L. Flaggs, "Knowledge Based Analysis and Design for Aerospace Structures," *ASME Special Publication*, AD-10, pp. 67-80, 1985.

INTEGRATED BUILDING DESIGN

M. Terk and S.J. Fenves
Carnegie Mellon University, Pittsburgh, PA, USA

ABSTRACT

The presentation is largely based on the enclosed paper, which will constitute Chapter 2 of a forthcoming monograph, *Concurrent Computer-Integrated Building Design* by S. Fenves, U. Flemming, C. Hendrickson, M. Maher, M. Terk, R. Quadrel and R. Woodbury (Prentice Hall, 1993). The presentation covers: motivation and history of the project; overview of the knowledge-based agents participating in the design process; overview of two successive system architectures for process, information and agent management; and a brief extrapolation to the future of integrated design environments.

1. Introduction

This paper describes the Integrated Building Design Environment project performed by the Engineering Design Research Center (EDRC), an NSF Engineering Research Center at Carnegie Mellon University.

1.1 Objectives of IBDE Project.

IBDE is a testbed for the exploration of integration and communication issues in the building industry. It integrates vertically the various design and planning tasks in the delivery of a constructed facility from the initial architectural programming for a building through structural and foundation system synthesis and design on to the planning of the construction activities.

The tasks involved are implemented as knowledge-based expert system agents. To the extent that they adequately capture the expertise relevant to their respective tasks, they can serve as surrogate experts, thus providing an environment where various experiments can be run to explore particular issues. In this way, conclusions reached can have a strong empirical basis.

To further emphasize the experimental nature of the project, IBDE was purposely designed to be modular, so that it can serve as a testbed for the empirical evaluation and calibration of integrated design support environments, a subject of intense research at EDRC and elsewhere. Experiments with these environments provide feedback to their developers and provides extrapolation to other design disciplines.

The domain of building construction was chosen as the subject of the exploratory study for two reasons. First, it represents the confluence of expertise and interests of faculty members in the Architecture and Civil Engineering Departments at Carnegie Mellon University, thus providing the basis for the type of collaborative, interdisciplinary research that EDRC fosters and that is difficult, if not impossible, to achieve in a hierarchically structured academic environment. Second, the building industry is fragmented into many diverse organizations, each responsible for only a portion of the overall building delivery process. Thus, only a multi-disciplinary research project can abstract from the "real world" situation and investigate the integration and communication issues without regard for organizational boundaries.

The decision to cast the project in the framework of computer-based integration and communication was similarly influenced by two factors. First, computer-based methodologies provide an ideal environment for experimental research in design and a suitable mechanism for eventually transferring research results into design practice. Second, the building industry, having achieved significant "islands of automation", is poised to take advantage of emerging computer-based technologies and make major steps in computer-based integration. In this respect, the project can serve as an early precursor in the investigation of integration issues to be faced by the building construction community as a whole.

At the outset, it is important to emphasize two significant limitations. First, IBDE was not intended to be, and is not, a prototype of a possible commercial building design system. IBDE was conceived from the beginning as purely an experimental, empirical testbed for the exploration of a host of issues in integration and communication. If the building design community judges IBDE to be adequate in addressing some of these issues, then the findings of the project can serve as one of the inputs to the functional specification of a prototype commercial system. It is premature to speculate what such a prototype, or an eventual "production" system would look like. The project's aim is purely to investigate the technical aspects of communication and integration that such a system would have to address. The term "technical" is to be emphasized, as the project did not deal with any of the organizational, jurisdictional, social and legal issues that are inherent in the present organizational structure of the building industry (for a discussion of these issues, see [2]).

As a second limitation, IBDE was not intended to serve as a normative, prescriptive model of how building design "ought to be done." Rather, in the empirical spirit characterizing the entire project, it was intended to provide a means for arriving at generalizations about the design process based on the experience and insights gained from experiments with the system.

1.2 The Design Process.

A successful engineering design project is a testament to teamwork and cooperation. Invariably, a complex design entails the participation of numerous design disciplines, each contributing a particular body of knowledge and expertise to the overall effort. If the project is large, each individual discipline will be represented by several design professionals, each of whom brings a unique set of talents and experiences to bear on the task at hand. The goal of integrated design is to bring out of this rich diversity of specialized knowledge and unique perspectives a result which achieves global design objectives and stands as a harmonious whole.

Design success is defined by the attainment of three goals:

- The resulting design must be *feasible*; that is, the contributions of the individual participants must be consistent and compatible with each other and with any externally-defined constraints on the solution (such as code restrictions or limits on budget and schedule).

- The resulting design must be *effective* with respect to the global objectives of the project. In this context, effectiveness may be considered as a relative measure for comparing the merits of alternate design solutions with respect to a defined set of objectives. Typical objectives include minimization of constructed cost, maximization of aesthetic appeal, etc.

- The process by which a feasible and effective design is achieved must itself be *efficient* [6]. An efficient design process is difficult to measure but easy to recognize. Efficiency does not imply that decisions are not revised and iteration does not occur, but that when revisions and iterations occur they can be responded to in an efficient and harmonious way.

Chronologically, IBDE addressed vertical integration first, in order to establish initial feasibility. The system and its component processes were then expanded in order to address issues of effectiveness and efficiencies.

1.3 Project History.

The IBDE project began shortly after the establishment of EDRC in 1986. It was originally conceived as an EDRC pilot project in which elements of all three of EDRC's major thrust areas (design for manufacturability, synthesis and design systems) would be brought to bear on a specific domain or problem area, namely, the design and construction planning of buildings. Subsequently, the project was organizationally located in the Design Systems Laboratory of EDRC, although strong ties to concerns for manufacturability (i.e., constructibility) and to synthesis methodologies have continued.

The history of the project can be concisely summarized into three phases:

- The initial phase (1986-87) involved the development of new knowledge-based agents and the adaptation of pre-existing agents, together with the conceptual design of the overall system architecture.

- The second phase (1987-89) was devoted to the initial integration of the agents to formalize the overall organization of the common representation of the project datastore and to the development of a preliminary version of the system architecture, with the first demonstration of the integrated system accomplished in August 1988.

- The third phase (1989-91) involved the design and implementation of a new system architecture, called the Integrated Facility Development Framework (IFDF), and the re-implementation of IBDE in the new framework.

In the two latter phases, a number of related projects were undertaken the results of which are presented in this paper but which were not physically incorporated into the IBDE system, largely due to the constraints and time pressures of graduate thesis research.

2. Overview of Agents

This section provides concise descriptions of the individual agents comprising IBDE. The agents in IBDE are classified into two groups: generators and critics. The term *generator* refers to a computer program, typically a knowledge based system in IBDE, that contributes to the development of the emerging design description. Each agent in IBDE is an independent computer tool that can execute outside of IBDE. One of the design objectives of IBDE was that the incorporation of the various agents into IBDE require little change to the original versions. The term *critic* refers to a computer program that does not contribute directly to the design description but evaluates the current description (possibly partial) and makes redesign recommendations.

The generators are:
- ARCHPLAN develops the building design concept

- CORE generates layouts of the service core

- STRYPES configures the structural systems

- STANLAY generates the layout and preliminary analysis of the structural system;

- FOOTER synthesizes and designs the building substructure;

- SPEX performs the design of structural components; and

- CONSTRUCTION PLANEX performs construction planning.

The critics are:
- CONSTRUCTION CRITIC provides constructibility criticism; and

- STRUCTURAL CRITIC performs a structural evaluation.

Each agent and critic is described according to the role it plays in the overall project development. As an information processing unit, each agent is first described in terms of its principal inputs and outputs. Since the agents are more than "black boxes", reasoning about various design and planning decisions, the agents are further described by the problem solving paradigm they use and how they transform a set of requirements (input) to a solution (output). Each agent is itself the result of research in computer supported building design.

ARCHPLAN. ARCHPLAN is a knowledge-based ARCHitectural PLANning expert system for the interactive development of a design concept. The input describes the given site, the client's program and budget, and applicable geometric constraints. The output provides three-dimensional information about the building's overall shape; the distribution of functions within the building, and the space allocated to the circulation system.

ARCHPLAN'S basic paradigm is *prototype refinement.* The program starts with a generic prototype of a typical office building, which is then refined by the user in interaction with the program and heuristic knowledge built into the program. The result is an instantiation of the prototype with the parameters (such as size and location of the building's footprint, number of floors, etc.) satisfying site- and problem-specific constraints and user-defined preferences. Prototype refinement takes place in three distinct, but interrelated decision modules. The *Site, cost, and massing module* (SCM) develops a massing model that will fit the given site and budget and a range of other parameters. Cost, site and massing options are treated as inter-dependent concerns. Conflicts are resolved based on the *Function module.* This module assists in determining the vertical and horizontal distribution of functions (office, retail, atrium, mechanical systems and parking) within the volume established by the previous module. The module proposes a three-dimensional layout scheme and presents it as solid or wire frame display. If conflicts occur with input data or earlier decisions, the program backtracks to the SCM module. The *circulation module* generates circulation proposals based on combinations of internal or external vertical circulation elements.

CORE. CORE generates layouts of the elements in the service core of the building (elevators, elevator lobbies, restrooms, emergency stairs, utility rooms, etc.). The input to CORE describes the overall geometry of the building and the expected size and location of the service core assigned by ARCHPLAN. CORE's output includes the number of elevator banks, the number and speed of the cars in each bank, the floors served by each bank and the layout of the banks, lobbies and other elements in the building core.

The spatial layout of the core is performed by an adaptation of LOOS, a general system for the generation of layouts in various domains. LOOS places particular emphasis on the generation of layout alternatives with interesting trade-offs. The major components of the LOOS architecture are a domain-independent *generator* able to generate layout alternatives; a domain-dependent *tester* able to evaluate the layouts produced by the generator, and a *controller* that steers the generation process into promising directions based on the test results. The paradigm employed is a form of *hierarchical generate-and-test* in which intermediate solutions are evaluated and, possibly, pruned from the search. The generator starts from an initial state and recursively expands it by adding one object at a time in all geometrically possibly ways, thus producing intermediate solutions that are immediately evaluated by the tester. The controller selects candidates for expansion based on these evaluations.

CORE first computes the optimal banking arrangement for the elevators and determines the needed auxiliary spaces. It then generates: layout alternatives for the first floor, where all elevator banks are present; layout alternatives for each zone; and compatible combinations of layouts.

STRYPES. STRYPES is a knowledge-based expert system that configures a structural system. It is based on the knowledge acquired through the development of HI-RISE [3]. The input to STRYPES includes: (1) the structural grid produced by ARCHPLAN, specifying potential locations for structural systems; (2) functional information about the building, such as intended occupancy and location and size of the service core; and (3) load information. The output of STRYPES provides the types and materials for the lateral (wind) and horizontal (gravity) load resisting systems.

STRYPES is implemented in EDESYN, an expert system shell for design synthesis in which the design knowledge base is represented as decompositions and constraints [4]. The decomposition knowledge comprises a taxonomy of systems and components, both as templates for representing specific designs and as alternatives for generating multiple solutions. Its *hierarchical decomposition paradigm* generalizes decomposition schemes that have been proposed elsewhere for building design under names such as "morphological box" or "analysis of interconnected decision areas". In EDESYN, the synthesis process is a constraint directed search through the taxonomy of systems and components in the knowledge base. Constraints are used to prune the search space at various levels of abstraction. As each subsystem or component is selected, it is checked for feasibility. An alternative is feasible if it is not eliminated by a constraint.

STRYPES generates alternative structural systems for resisting lateral and gravity load. Lateral system alternatives considered are 3-D systems (core system) and orthogonal 2-D systems, selected from rigid frames, braced frames or shear walls. For gravity systems, alternate 2-D horizontal subsystems and vertical supports are considered.

STANLAY. STANLAY, also developed using EDESYN, performs two major tasks for the preliminary structural design of the building. The first task is the layout of the structural system specified by STRYPES, the second is an approximate analysis of the structural system. The input to STANLAY includes: (1) the structural grid; (2) the architectural function of the building; and (3) the structural systems selected by STRYPES. The output of STANLAY is the location of the lateral and gravity load systems, the approximate load effects on the structural components and the grouping of the components.

The layout task involves identifying several possible locations of the lateral load system and specifying the location of the gravity load system. The location of the lateral load systems requires the specification of 2D vertical subsystems, such as rigid frames or shear walls, and their location on the grid. Based on the layout and location, loads are distributed to the components and component load effects are determined using approximate analysis techniques.

FOOTER. FOOTER is an expert system that performs the preliminary design of the foundation components of the building; it is also implemented in EDESYN. The input to FOOTER includes: (1) soil conditions, such as the presence of obstruction, location of water table, depth of bedrock, and soil classification; and (2) imposed loads provided by STANLAY. The output of FOOTER is a description of a footing or pile for each column and/or shear wall.

The foundation design task is decomposed into the following subtasks: building characterization, site characterization, and foundation synthesis. The building and site characterization tasks use the relevant view of the project data store and infer or add data relevant to the design of foundations. This task is similar to the task performed by foundation engineers in determining the load requirements of the building and the bearing capacity of the soil. Foundations are synthesized from combinations of shallow, compensated, or deep foundations. Each category of foundation types is decomposed into the relevant design parameters. Alternative values for the parameters and constraints on their application are considered by FOOTER to design feasible alternatives.

SPEX. SPEX is a knowledge-based system for structural component design. It is responsible for the preliminary design of structural components for the structural system configured by STANLAY. SPEX receives as input the design parameters for each component group: (1) type of component (e.g., beam, column); (2) length; (3) material (steel or concrete); and (4) estimated load effects on the component. The SPEX interface supplies the material grade, the name of the design standard, the design focus, and an optimality criterion. The output of SPEX is the description of the optimal component.

SPEX implements a heuristic *generate-and-test* design strategy in which components are designed by applying three types of knowledge: knowledge contained in design standards; "textbook" knowledge of structural, material and geometric relationships; and designer-dependent design expertise. Design expertise is represented in SPEX by rules which express the designer's *intent* for the structural performance of the component. This intent serves as the *focus* for retrieving the relevant provisions from the applicable design standard. SPEX generates a *trial* design by optimizing the structural component with respect to the constraints corresponding to the design focus. This trial design is then checked against *all* constraints applicable to the component. If the trial is unacceptable (i.e., it violates some constraint) a new focus is formed and the process repeated. In the IBDE implementation, no iteration is performed, since only a preliminary design is sought.

The subtasks of SPEX are: retrieve design focus; retrieve design standard requirements corresponding to focus; generate constraints resulting from the requirements; and select optimal component satisfying the constraints.

CONSTRUCTION PLANEX. CONSTRUCTION PLANEX is a knowledge-based expert system to assist the construction planner. It is responsible for estimating the basic attributes of cost, construction resource requirements and time schedule to construct the building. The input to PLANEX consists of: (1) specifications of the physical components of the structure and foundation provided by the other agents; (2) site information (such as soil type and elevations); and (3) resource availability (such as number of crews or equipment types). The output from PLANEX consists of a complete plan of construction activities including a provisional schedule and cost estimate.

CONSTRUCTION PLANEX is an implementation of PLANEX, which is primarily based on the *nonlinear planning paradigm* developed in Artificial Intelligence in which elements of the initial partial plan are expanded, retracted and re-expanded until the sequence of activities meets all precedence and resource requirements. PLANEX first decomposes each structural or foundation component into element activities (e.g. formwork erection, reinforcement placement, concrete placement, curing and formwork stripping for a concrete component) and aggregates element activities into project activities (e.g. formwork erection on floor 10). PLANEX next selects construction technologies for each activity based on its knowledge about appropriate alternative technologies (e.g. place concrete by pumping vs. lifting buckets by crane). PLANEX will reject design elements for which construction knowledge or necessary equipment are not available. Finally, PLANEX determines the resources required in terms of crews and equipment and produces the cost, resource requirements and time schedule estimates.

The results of PLANEX (the project cost and duration estimates, the equipment and resource requirement profiles during the construction phase, and the element and project activities by number and kind) may be collectively viewed as critiques of the design produced by the other processes.

CONSTRUCTION CRITIC.[1] The construction critic is a rule-based system for evaluating the constructibility of a particular design. The prototype system was restricted to diagnosis and criticism of steel framing systems. The critic examines issues such as excessive length and weight of individual elements and the variability in the numbers and types of individual beam elements. The system applies a series of heuristic rules to a particular design and generates a series of specific criticisms and an overall "constructibility score" for a design.

STRUCTURAL CRITIC.[2] The structural critic is a knowledge-based interface between IBDE and a commercial finite element analysis program. The input to the critic consists of: (1) loads acting on the building; (2) the structural configuration determined by STRYPES and STANLAY; (3) the approximate member load effects computed by STANLAY; and (4) the structural component properties determined by SPEX. The output of the critic is a set of messages identifying: (1) excessive discrepancies between the estimated load effects and the load effects determined by analysis; (2) poor grouping of elements; and (3) excessive lateral deflection or drift.

The structural critic is essentially a diagnostic expert system utilizing the *heuristic classification* paradigm to classify the structural system with respect to the three categories of dysfunctions listed above. It consists of four modules: (1) an input processor; (2) a pre-processor or modeler for generating the analysis model; (3) the analyzer proper; and (4) a post-processor or interpreter that performs the comparisons and generates the output messages.

3. Overview of System

This section provides a concise description of the IBDE system that integrates the actions of the individual agents.

3.1 Initial version of IBDE
The architecture of the initial version of IBDE consisted of six major components [1]. The system components and their function are:

- the *controller* is responsible for activating the agents and communicating project information between them;

- the *status blackboard* records the status of the processes;

- the *datastore manager* is responsible for retrieving, storing, and translating data for individual agents;

- the *project datastore* records the global representation of the project information;

- the *common user interface* is a graphical and textural display of the project information and the current status of the agents; and

- the *tool set* consists of the initial agents described in Figure 2.1.

---

[1]The prototype version of the construction critic [5] has not been directly incorporated into the IBDE environment.

[2]The prototype version of the structural critic [7] has not been directly incorporated in the IBDE environment.

The architecture of the initial version is illustrated in Figure 1.

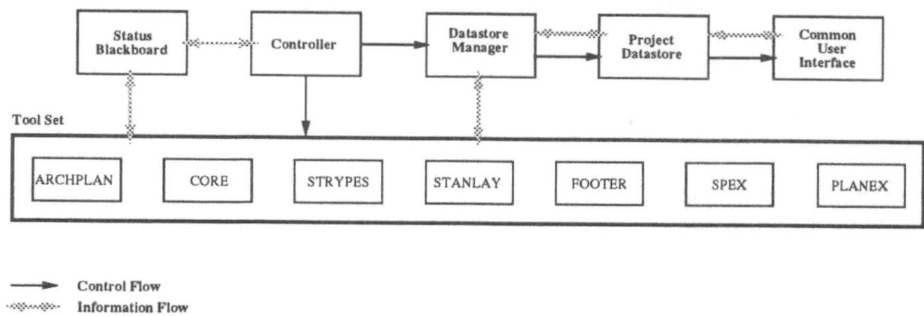

Figure 1: Initial Version of IBDE

The function of the system components is briefly described below.

Controller. The controller is responsible for activating the individual agents. Each agent is in one of three states: *pending*, *active*, or *completed*. Whenever an active agent terminates, it sends a new status message which the controller posts on the status blackboard. The message also signifies whether the agent was successful in producing a feasible solution or not.

The initial implementation provides a very limited control strategy, namely, event-driven, sequential agent activation. The controller maintains only the following static description about each agent process: (1) preconditions for its execution, namely, the agent(s) that must have been successfully completed before the current agent can be activated; and (2) machine on which the process runs. When the preconditions of a process are satisfied, the controller causes that process to be activated.

The controller is implemented on top of the DPSK (Distributed Problem Solving Kernel) system developed at CMU [1]. DPSK provides an environment for distributed problem solving on multiple machines by programs written in several languages. DPSK provides utilities for sending messages and signals between processes running on different machines, generating and responding to events, and communicating between processes by means of a Shared Memory accessible to all the processes. DPSK was designed to facilitate the implementation of a variety of cooperative problem-solving architectures; the initial IBDE implementation, with fixed precedence ordering between processes, is a relatively simple application of DPSK.

Status Blackboard. The status blackboard records the status of processes active in the environment and provides a medium for communication between the controller and the agents in the tool set.   The status blackboard is organized as a distributed shared memory and is implemented using the utilities provided by DPSK. The information on the status blackboards indicates the processing status of all agents in the tool set of IBDE.

Datastore Manager. The datastore manager works in concert with the controller and is responsible for supplying the input data to the agents and retrieving their output data. Prior to initiating an agent by the controller, the datastore manager transfers the input data to the machine on which the agent resides. When an agent terminates, it leaves its output on its own machine;

when its termination message is received, the controller causes the datastore manager to retrieve the data from the agent's machine and merge it into the datastore.

The datastore manager is responsible for generating views or subschemas as needed by the processes, including all format and structural conversions. The local views of all of the agents consist of sets of objects with attributes in the respective implementation languages of the agents. Furthermore, in most agents, no explicit distinction is made between input and output attributes; the object contains all necessary attributes.

In the initial implementation, data is communicated between the processes by means of files. Each file contains all instances of a particular object type (e.g., beams or columns). There is a one-to-one correspondence between the objects in the files and the individual local objects, although there are differences in format and attribute names. The datastore manager is responsible for format and name translation as well as transferring the appropriate files to and from the agents.

Project Datastore. The project datastore holds the global representation of the building and serves as the repository of data communicated between the IBDE agents.

The datastore is hierarchically organized as a tree of related objects. Objects may represent very high-level abstractions, such as the entire building, or very detailed information, such as individual building elements. The hierarchy primarily represents part-of relations, where each object is a component of a higher-level parent object. Provisions are also made for representing is-alternative relations, where an object is an alternate design solution of the parent object. Through this latter relation, redesign in response to critiques received is readily supported. This overall organization is independent of the internal global schema implementation. With this organization, each agent can access the contents of the datastore relevant to it, but not of the segments relevant to the other agents. This organization has supported the concurrent development of the agents and provides complete data and process independence among the agents. The datastore provides at all times a complete snapshot of the current state of the building design and construction planning process.

Common Display Interface. The data residing in the datastore is inaccessible to users without a common user display interface. As the interface is intended for a variety of users with different backgrounds, it must conform to certain graphical standards and should exhibit a degree of intelligence. An interface of this type was developed for the IBDE project. It provides a uniform set of interface facilities for the following functions:

- *Graphical display of the status of agents.* Each agent is shown as either pending, active, or completed.

- *Graphical display of data at any level of the project datastore representation.* As soon as an agent is completed, the content of the datastore can be displayed. The user sees the geometric representation of this data as three-dimensional objects or as charts and symbols.

- *Textual and graphical display of object classes.* The user selects one of the datastore objects directly from a menu, and the geometric and textual information is displayed.

- *Graphical display of selected items.* The designer can specify constraints to view objects of a certain class or that fall within user-defined limits. All objects found conforming to the constraints are highlighted on the graphical display.

- *Graphical navigation to select specific objects.* Once selected graphically, the object is highlighted and the appropriate datastore object appears on screen in a pop-up window.

Critique. The initial version of IBDE is an example of a design environment using a *tool-centered* approach: it uses a controller that executes a fixed set of tools (agents), thereby tightly coupling the functions of the design environment and the tools it manages, as illustrated in Figure 2.2. This tight coupling between the *strategy level* comprising the system components and the *resource level* comprising the tools (agents) may be acceptable in application areas with well-established, static collections of agents and agreed-upon problem-solving strategies. Even though the system components were designed with flexibility and change in mind, the tool-centered approach turned out to be unacceptably restrictive for the exploratory environment of the IBDE project: additions of new agents and critics, and changes in the process, information and agent management strategies encountered major obstacles. Consequently, a new approach was developed.

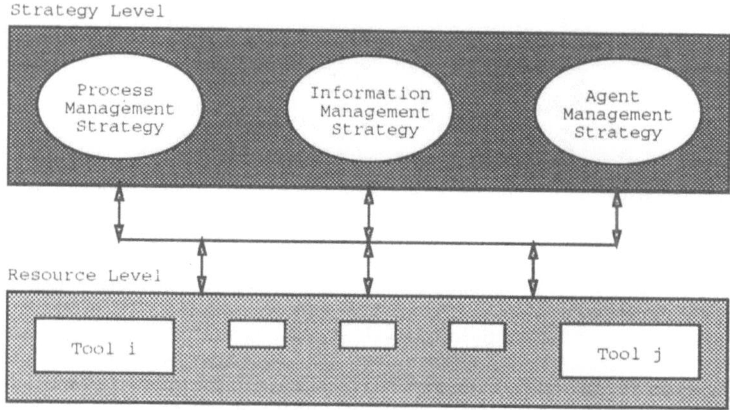

Figure 2: Tool-Centered Approach

3.2 IBDE-2.

IBDE-2 was created using the Integrated Facility Development Framework (IFDF), developed as part of the IBDE project. IFDF is based on a *problem-centered* approach.

Problem-Centered Approach. The problem-centered approach decouples the strategy and resource levels through the use of tool-independent representations of the building and of the building development process. Consequently, the design environment provides a mapping between the strategy level and the representation level as well as between the representation level and the resource level. The tool-independent representations provide a level of separation between the problem solving knowledge (*how* to manage the various facets of building development) and the resources available to solve the problem (*what tools* are are available to solve and manage the sub-problems). Figure 2.3 shows the conceptual organization of this approach.

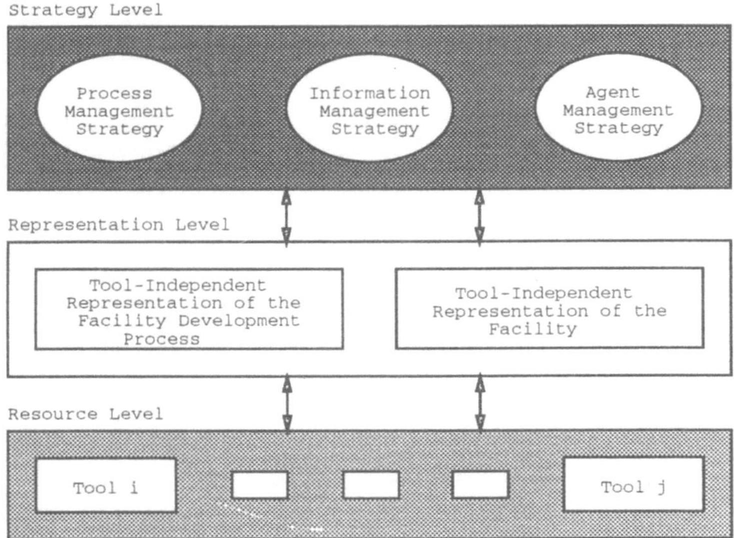

Figure 3: Problem-Centered Approach

The Representation Level of IFDF consists of two representation: the *task-aspect* representation of the building development process; and the *facility description schema* representation of the building. The task-aspect representation provides IFDF with a tool-independent representation of the building development process. The task-aspect representation consists of two primitives: a *task* that represents an activity performed during building development; and an *aspect* that represents a collection of information used or generated by individual tasks. Both tasks and aspects are hierarchical in nature. The notion of aspects, as defined by the task-aspect representation, names the information used in building development without describing its semantic structure. Therefore, the aspect hierarchy, by itself, can not provide a tool-independent representation of the building. This representation is provided by the *facility description schema* of IFDF. The facility description schema established in IFDF, uses the notions of *objects* and *links* as basic units of representation. Objects represent a group of data that can be manipulated as a single entity while links represent relationships between objects. The two representations used by IFDF are linked by a two-way mapping between the aspects in the task-aspect representation and the objects in the facility description schema. This mapping reflects the realization that a number of information management issues, such as versioning and consistency enforcement, are influenced by the building development process.

Architecture of IFDF. The architecture of IFDF consists of six major components, shown in Figure 2.4. The Strategy Level consist of the following components:

- The *Development Process Manager* establishes the process management strategy of the design environment and manages the mapping between the strategy and the task-aspect representation of the building development process;

- The *Data Exchange Manager* establishes the information management strategy of the design environment and manages the mapping between the strategy and the facility description schema representation of the building;

- The *Activator Set* establishes the agent's interactions with all other components of the design environments and maps the functionality and information requirements of an agent into the task-aspect representation of the development process.

The Resource Level consist of the following components:

- *Development Process Description.* The Development Process Manager provides storage for information relating to the process management issues addressed by the environment;

- The *Tool Set* transforms stand-alone computer tools into agents capable of functioning within the design environment;

- The *Facility Description* provides storage for the information used by the design environment; and

- The *Global User Interface*(GUI) displays information of the Facility Description.

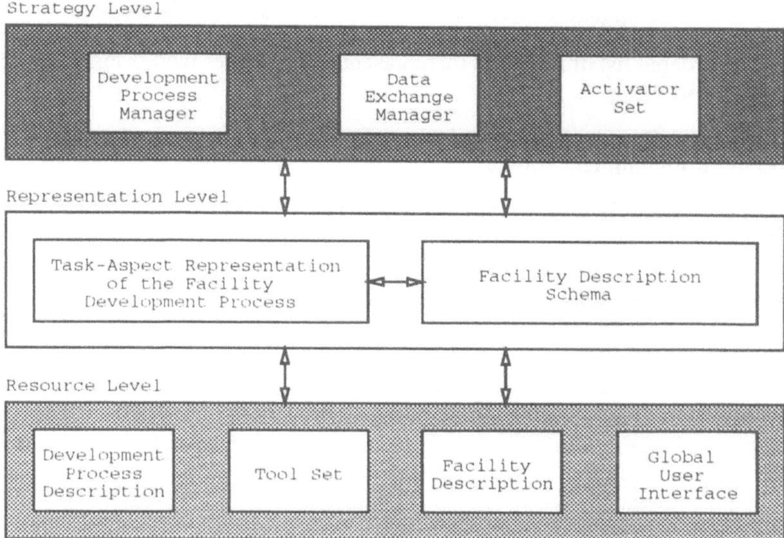

Figure 4: IFDF Architecture

The components of IFDF, when combined with a set of agents, create design environments addressing the building development process.

The function of the IFDF system components is briefly described below.

Development Process Manager. The Development Process Manager is responsible for organizing and managing the problem-solving activities of the agent in a design environment. The Development Process Manager is designed to aid the user in defining the problem as a set tasks and in managing the execution of these tasks by the agents in the design environment.

The task-aspect representation of the building development process used by IFDF requires a design environment to decompose a problem into a set of tasks that have to be achieved to solve the overall problem. The process of obtaining a problem decomposition is posed as a search for a set of activities (tasks) that map a set of available aspects (the context) into a set of aspects representing the desired state of building development (the goal). The search space for this planning problem consists of the aspect hierarchy defined by the task-aspect representation of the building development process and the operator space is comprised of the tasks in the task hierarchy. The Development Process Manager provides a hierarchical planner capable of generating an ordered set of tasks that map between the context aspects and the goal aspects.

Once the problem is decomposed into an ordered set of tasks, the Development Process Manager is responsible for managing the execution of these tasks towards an efficient solution of the overall problems. At present, the responsibilities of the Development Process Manager can be separated into two major categories: task distribution and contingency resolution. Task distribution involves matching the tasks in the problem decomposition to agents best capable of processing then. IFDF uses the contract net protocol to implement the market approach to task distribution. The Development Process Manager acts as a manager in the contract negotiation by generating a contract for a task, broadcasting it to all interested agents, and collecting and ranking the replies. The ordered list of replies is then presented to the user who is responsible for selecting the most appropriate agent.

The Development Process Manager also monitors the following four types of contingencies that may occur during problem-solving:

- Allocation Failure. No agents in the environment is capable of solving a task in the problem partition.

- Hardware Failure. An agent has failed to perform a task because of an occurrence external to its processing (e.g. communication failure).

- Agent Failure. An agent fails to obtain a solution to a task it has been assigned.

- Agent Critique. An agent is able to find a solution to its task but also has suggestions about how the solution can be improved in the next design iteration.

The Development Process Manager contains a set of agent-independent strategies for resolving each of these contingencies.

Data Exchange Manager. The Data Exchange Manager is responsible for supplying the input data to the agents and retrieving their output data. Prior to the activation of an agent, the agent issues a data retrieval message that causes the Data Exchange Manager to retrieve the information in the input aspects associated with the task the agent is assigned from the Facility Description and to place it in the proper location on the machine on which the agent resides. Once the agent has finished processing, it issues a data storage message to the Data Exchange Manager. As the result of this message, the Data Exchange Manager retrieves information in the output aspects associated with the completed task from the output file on the agent's machine and stores it in the Facility Description.

The Data Exchange Manager is responsible for managing all semantic and syntactic translations required to map between the local representations used by the individual agents and the representation used by the Facility Description. These translations are facilitated by a mapping between the aspects in the task-aspect hierarchy and the objects in the facility description schema used in IFDF and the mapping between the notion of an object in the facility description schema and the primitives in the representation used by the Facility Description. The Data Exchange Manager provides utilities used by design environments to establish and manage these mappings.

In this implementation, information is communicated between an agent and the Facility Description by means of files. These files contain information represented either using the agent's local representation, in which case they must be parsed by the Data Exchange Manager into operations on the Facility Description, or as direct commands to the utilities in the Data Exchange Manager that operate on the Facility Description.

Activator Set. The Activator Set component of IFDF provides an *activator* module for each agents in the tool set of the environment. The activator module is responsible for establishing the agent's capabilities in the problem-solving process and contains utilities that allow the agent to participate as a contractor in the task distribution stage of problem-solving in IFDF.

The activator module is implemented as a generic processor that obtains agent-specific knowledge through the information specified in a *tool-description* file. The tool-description file contains the information about the agent's name, classification (generator or critic), and physical location. In addition, the tool-description contains the conditions upon which the agent will activate. In the case of a generator agent, these conditions comprise of a list of tasks the agent is capable of and a set of conditions under which the agent will accept a contract for a task. In the case of an critic agent, the activation conditions consist of names of aspects the agent is interested in examining. The run-time behavior of an activator module varies based on whether an agent is classifies as a generator or a critic. An activator for a generator agent functions as a contractor in the contract-net task distribution in IFDF. The activator monitors the environment for contract announcement, formulates bid messages for contract it is capable of performing and, if chosen to execute a task, initiates and monitors the agent's processing. The activator for a critic agent initiates the agent's processing as soon as the aspects it is interested in become available.

Development Process Description. The Development Process Description provides storage for information communicated between components of IFDF during problem-solving. The Development Process Description is subdivided into two areas: the *contract data-space*, containing all information exchanged during task distribution; and *constraint data-space* containing all information used during contingency detection and resolution.

The Development Process Description is build on top of the Distributed Problem Solving Kernel (DPSK) and is implemented as a distributed shared memory. This distributed shared memory paradigm allows easy access to communications exchanged between the components of a design environment while reducing the communication bottleneck inherent in the global shared memory paradigm.

The contents of the Development Process Description can be used to review the current and previous stages of the building development process. In addition, the Development Process Description can serve as a central location for obtaining and evaluating information about the performance of various approaches to problem-solving in a design environment.

Tool Set. The Tool Set component of IFDF provides utilities used to convert the stand-alone tools in the tool set of IBDE into agents capable of functioning within the IFDF framework.

IFDF uses the *tool encapsulation* approach to convert stand-alone tools into agents capable of functioning within a design environment. Tool encapsulation involves surrounding a tool with modules that augments the tool's original functionality and allow it to participate in a design environment. The Tool Set of IFDF provides a set of pre- and post-processor modules that allow a stand-alone tool to participate in the problem-solving activities of a design environment.

Facility Description. The Facility Description component of IFDF stores and manager all information used by a design environment during problem solving. As a result, the contents of a Facility Description must match the information contents of the facility description schema of the design environment.

The current Facility Description is implemented on top of a commercial relational database system. The relational data model was selected because it has been shown effective for managing large amounts of data and because the current generation of commercial data management systems have a number of extensions that correct some of the shortcomings of the relational data model with respect to managing engineering data. The consistency of the Facility Description is ensured by not allowing agent to access it directly. All operations on the Facility Description are serialized through the Data Exchange Manager which is the only component of IFDF that is allowed to modify the Facility Description.

Global User Interface. The Global User Interface allows the user to graphically browse the current contents of the Facility Description. The functionality of the Global User Interface in IFDF is similar to the functionality provided by the the Common Display Interface component of IBDE.

IBDE-2 Implementation in IFDF. Figure 5 shows the task and aspect hierarchies established to represent the problem-solving activities in IBDE-2. Level 0 of the task hierarchy consists of four tasks while level 1 describes the sub-tasks produced by decomposing the StructSysDevelopment task. The decomposition of the StructSysDevelopment task creates two aspect hierarchies: the ArchProgram aspect is decomposed into the BuildingDescription aspect; and the GridDescription, StructSysType, StructComponentLoc and StructComponentDesign aspects are composed to produce StructSysDescription aspect. This representation of the building development process is used to implement the process management strategy in the environment and is used by the Development Process Manager of IFDF to manage the problem solving process.

The facility description schema of IBDE is organized as a hierarchy of objects. At the present time, the facility description schema in IBDE-2 supports only one type of links between objects: the "part-of" links. Figure 6 shows the top level organization of the facility description schema in IBDE-2.

Each tool in the tool-set is classified into one of three categories based on their contribution to the building development process. The tool in the tool-set of IBDE-2 are classified as follows:

- Generators: ARCHPLAN, GRID[3], STRYPES, STANLAY, SPEX, FOOTER and PLANEX;

---

[3]GRID was introduced into IBDE-2 to allow the user to generate structural grids that override the grids generated based on the limited structural heuristics of ARCHPLAN.

- Critics: STRUCTURAL CRITIC and CONSTRUCTION CRITIC[4]; and

- Utility: GUI; the Graphical User Interface.

The mapping between the elements of the task-aspect representation of the building development process and the processing capabilities of these tools is as follows:

- ARCHPLAN: performs ProgramDevelopment task;

- GRID: performs StructGridSelection task;

- STRYPES: performs StructSysSelection task;

- STANLAY: performs StructSysDesign task;

- SPEX: performs ComponentDevelopment task;

- FOOTER: performs FoundationDevelopment task;

- PLANEX: performs ConstructionPlanDevelopment task;

- GUI: displays ArchProgram, StructSysDesign, StructSysDescription and FoundationDescription aspects.

Figure 7 show the architecture of IBDE-2.

---

[4]The STRUCTURAL CRITIC and CONTRUCTION CRITIC have not been included in IBDE-2.

Level 0

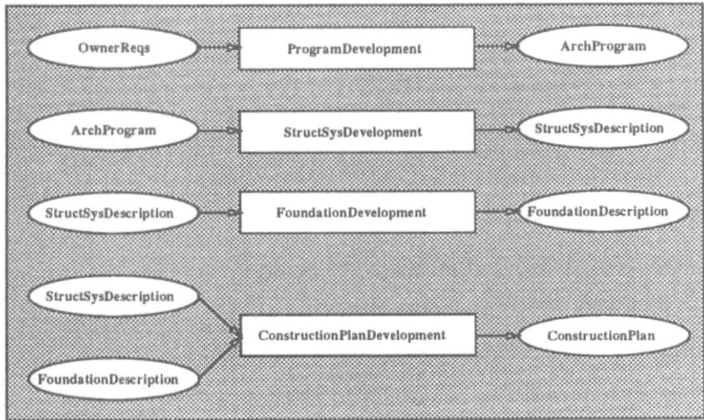

Level 1: StructSysDevelopment

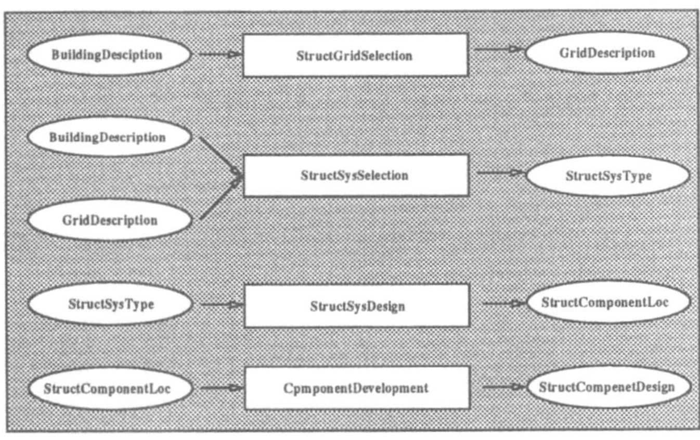

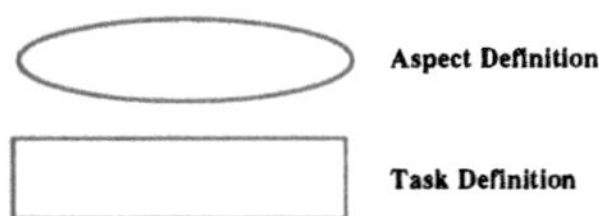

Figure 5: Task-Aspect Representation of IBDE-2

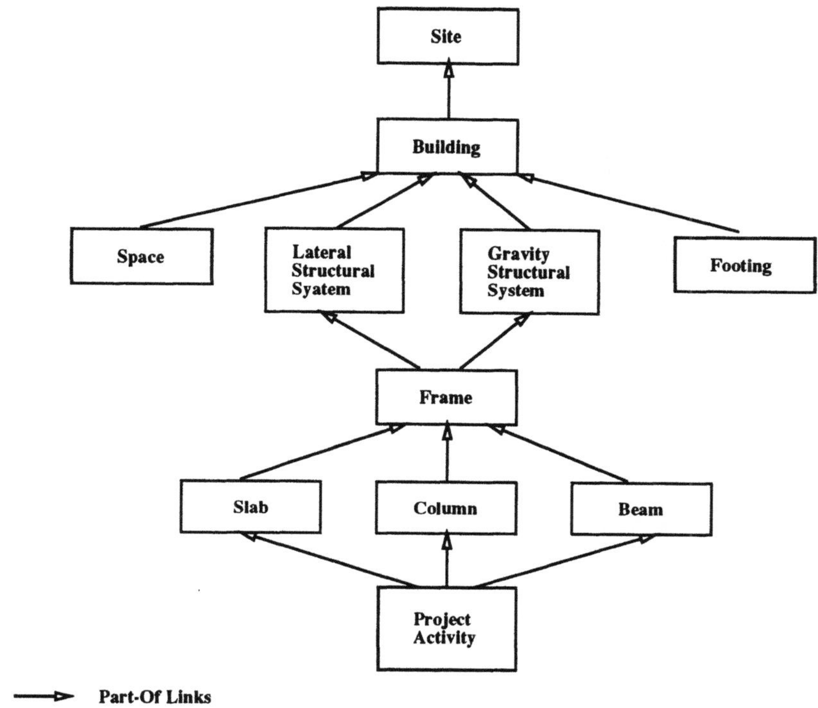

Figure 6: Facility Description Schema of IBDE-2

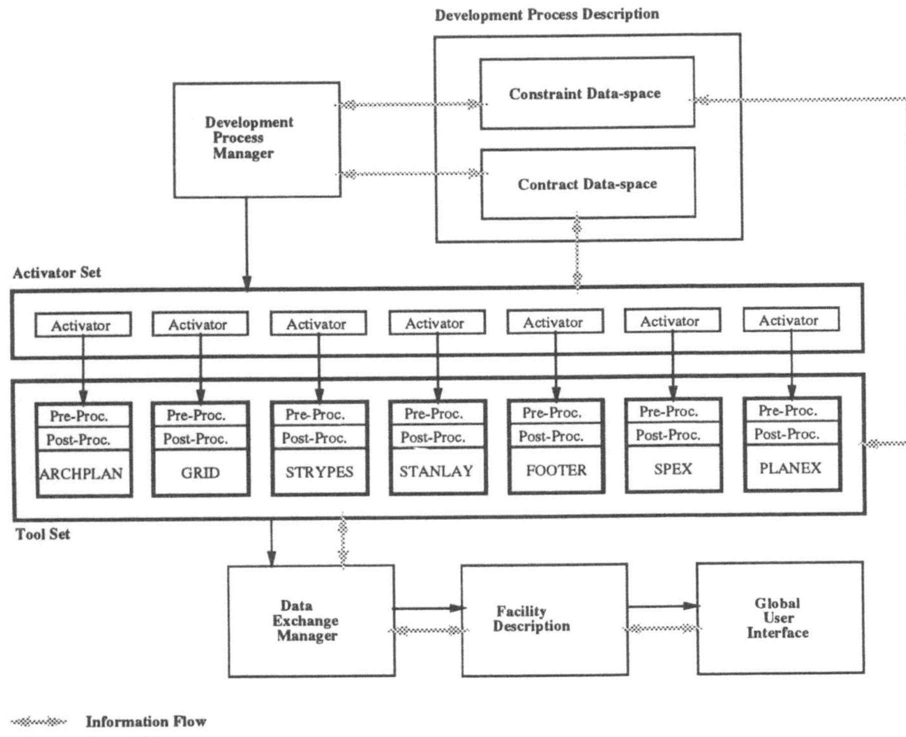

Figure 7: Architecture of IBDE-2

References

[1]   S. J. Fenves, U. Flemming, C. T. Hendrickson, M. L. Maher, and G. Schmitt.
        Integrated Software Environment for Building Design and Construction.
        Computer-Aided Engineering 22(1):27-36, 1989.

[2]   Hendrickson, C. and T. Au.
        Project Management for Construction.
        Prentice-Hall Ltd., 1989.

[3]   Maher, M.L., and Fenves, S.J.
        *HI-RISE: An Expert System For The Preliminary Structural Design Of High Rise
            Buildings.*
        Technical Report R-85-146, Carnegie Mellon University, November, 1984.

[4]   Maher, M.L.
        Engineering Design Synthesis: a Domain Independent Representation.
        Artificial Intelligence for Engineering Design 1(3):207-213, 1988.

[5]   Miller, E. M.
        Implementing Computer Aided Constructability Critics.
        Master's thesis, Department of Civil Engineering, Carnegie Mellon University, 1990.

[6]   Morse David V.
        Communication in Automated Interactive Engineering Design.
        PhD thesis, Carnegie Mellon University, 1990.

[7]   Priti Vora.
        A structural critic for the IBDE.
        Master's thesis, Department of Civil Engineering, Carnegie Mellon University, 1989.

COMPUTER AIDED COLLABORATIVE PRODUCT DEVELOPMENT

D. Sriram
Massachusetts Institute of Technology, Cambridge, MA, USA

1 Introduction

Engineering a product involves several stages (see Figure 1). In the first stage, a market survey for potential products is performed. This is followed by the conceptualization stage, where a product is conceived either as a result of a need or a potential profit motive (determined at the market survey stage). In the research and development stage, the information needed for the design of the product is developed. Design involves configuring the product based on several constraints. The manufacturing process yields the actual product. The product is then tested for quality in the testing stage and marketed in the marketing stage. The maintenance of the product is a service provided by most organizations. The above process is iterative (shown by bent arrows) and collaborative.

In traditional product development, the lack of proper collaboration between various engineering disciplines poses several problems, as expounded by the following clip from Business Week, April 30, 1990, p. 111 (see Figure 2 for a typical scenario in the AEC industry).

> "The present method of product development is like a relay race. The research or marketing department comes up with a product idea and hands it off to design. Design engineers craft a blueprint and a hand-built prototype. Then, they throw the design "over the wall" to manufacturing, where production engineers struggle to bring the blueprint to life. Often this proves so daunting that the blueprint has to be kicked back for revision, and the relay must be run again - and this can happen over and over. Once everything seems set, the purchasing department calls for bids on the necessary materials, parts, and factory equipment - stuff that can take months or even years to get. Worst of all, a design glitch may turn up after all these wheels are in motion. Then, everything grinds to a halt until yet another so-called engineering change order is made."

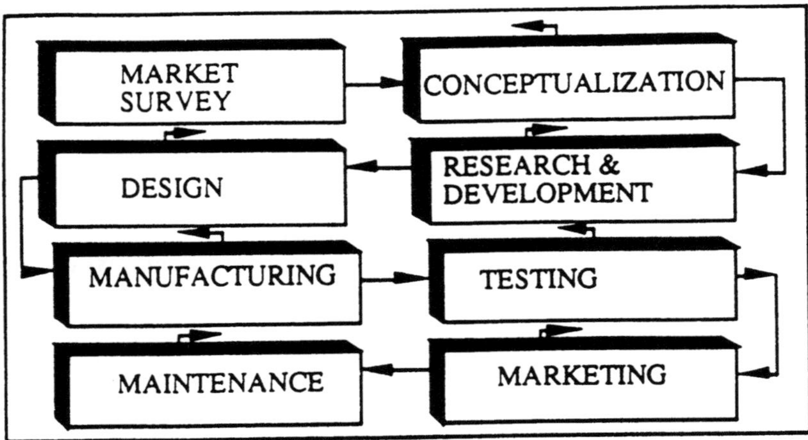

Figure 1: Engineering a Product

(Bent arrows indicate that the process is iterative)

Such problems routinely arise in the construction industry. Because designers find coordination among themselves difficult, they leave this task to construction managers or the contractor. Thus working drawings, used to inform the contractor of the product, lack detail. Shop or fabrication drawings are required from the contractor to document details, but potential conflicts among trades are often unrecognized until construction begins. Several undesirable effects are caused by this lack of coordination.

1. The construction process is slowed, work stops when a conflict is found.

2. Prefabrication opportunities are limited, because details must remain flexible.

3. Opportunities for automation are limited, because capital intensive high speed equipment is incompatible with work interruptions from field recognized conflicts.

4. Rework is rampant, because field recognized conflicts often require design and field changes.

5. Conservatism pervades design, because designers provide excessive slack in component interfaces to avoid conflict.

6. The industry is unprepared for the advent of automated construction, as the need for experience in design limits choice to available materials placed by hand.

All of these problems decrease productivity. In addition, failures, such as the Hyatt Regency collapse [37] which happened in July 1981, occur more often then they should.

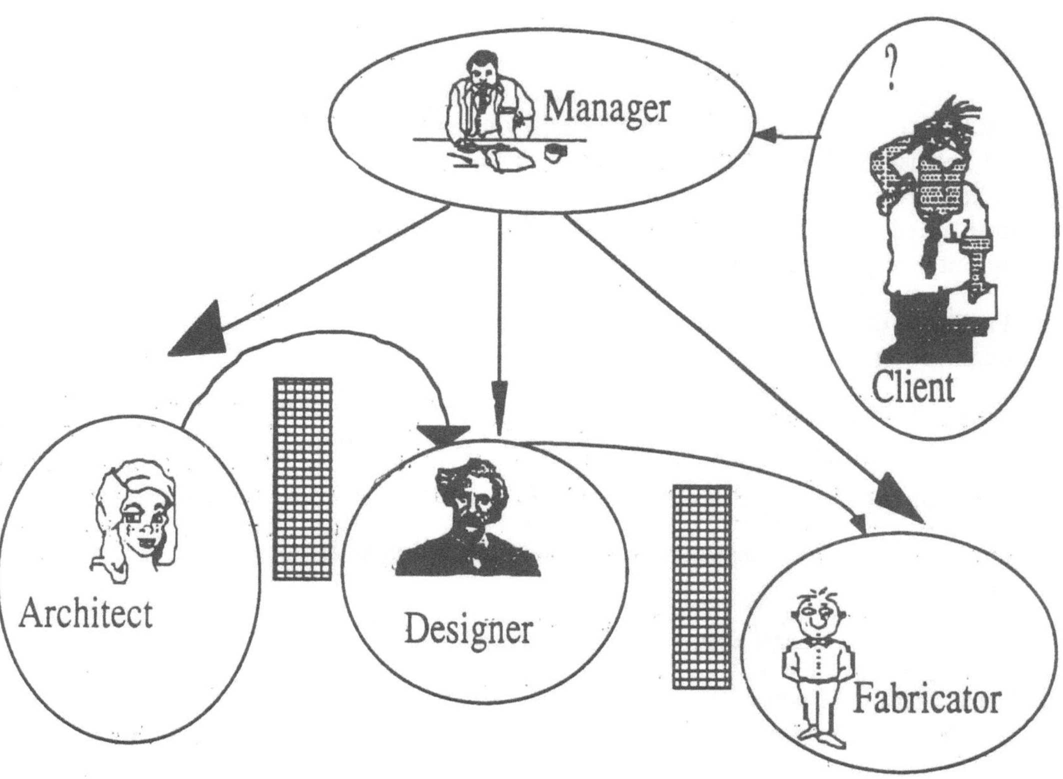

Figure 2: Over the Wall Engineering

Several companies have addressed the coordination problem by resorting to a more flexible methodology, which involves a collaborative effort during the entire life cycle of the product (See Figure 3). It is claimed (Business Week, April 30, 1990) that this approach[1] results in reduced development times, fewer engineering changes, and better overall quality. The importance of this approach has been recognized by the Department of Defense, which initiated a major effort - the DARPA Initiative in Concurrent Engineering - with funding in the millions of dollars.

It is conceivable that the current cost trends in computer hardware will make it possible for every engineer to have access to a high performance engineering workstation in the near future. Collaboration will be facilitated by a network of computers/users, as shown in Figure 4; we use the term *agent* to denote the combination of a human user and a computer. This is the philosophy that we have adapted in our approach, where we are developing computer aided tools - collectively called DICE (Distributed and Integrated environment for Computer-aided Engineering) - to address the following objectives:

1. Facilitate effective coordination and communication in various disciplines involved in engineering.

2. Capture the process by which individual designers make decisions, that is, what information was used, how it was used and what did it create.

3. Forecast the impact of design decisions on manufacturing or construction.

4. Provide designers interactively with detailed manufacturing process or construction planning.

5. Develop a few design agents for illustrating our approach.

In the next section we will outline the research issues that need to be addressed in computer-aided cooperative product development. This is followed by descriptions of the various projects we are pursuing under the DICE framework.

2 Research Problems

A computer-aided cooperative product development environment would involve a close collaboration between computer scientists, engineers, cognitive scientists, and management personnel. We believe that the following research areas will need to be addressed (see Figure 5).

---

[1]The terms Concurrent Engineering, Collaborative Product Development, Cooperative Product Development, Integrated Product Development and Simultaneous Engineering are often used to connote this approach.

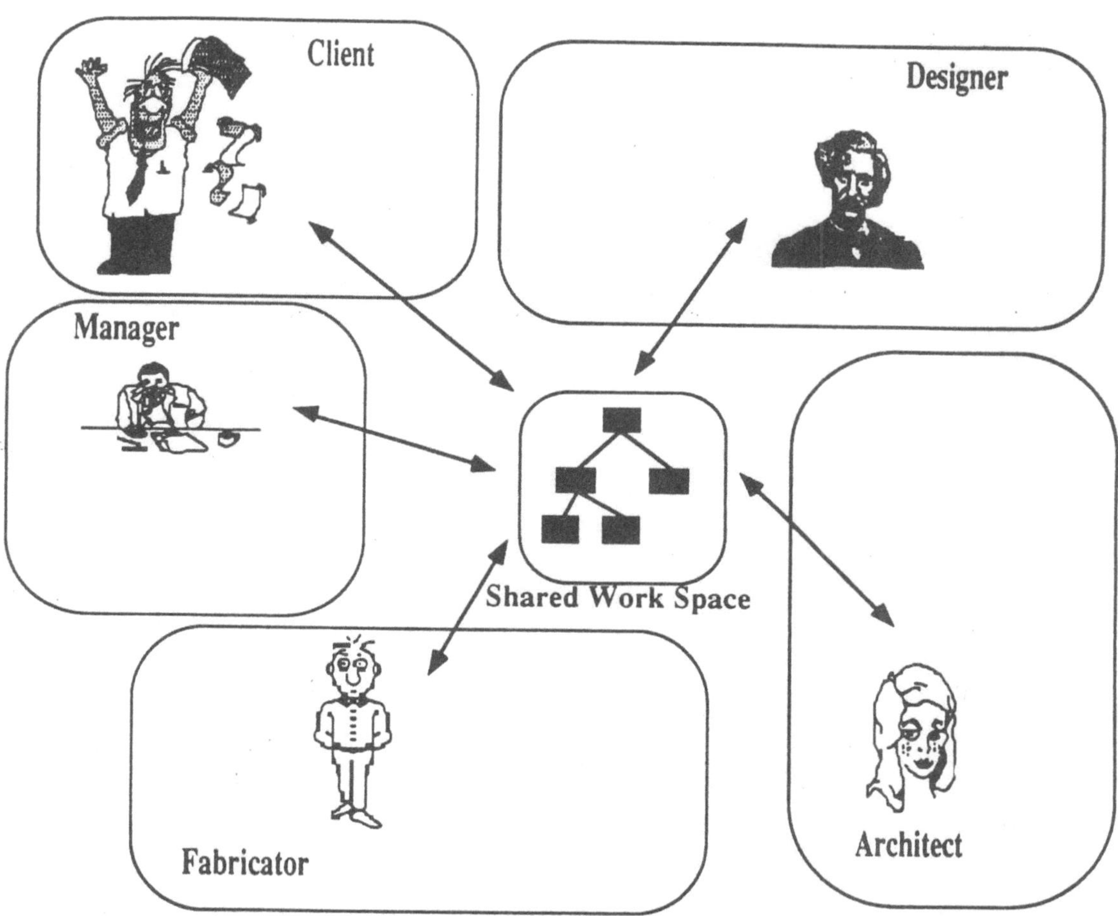

Figure 3: Modern view of Product Development

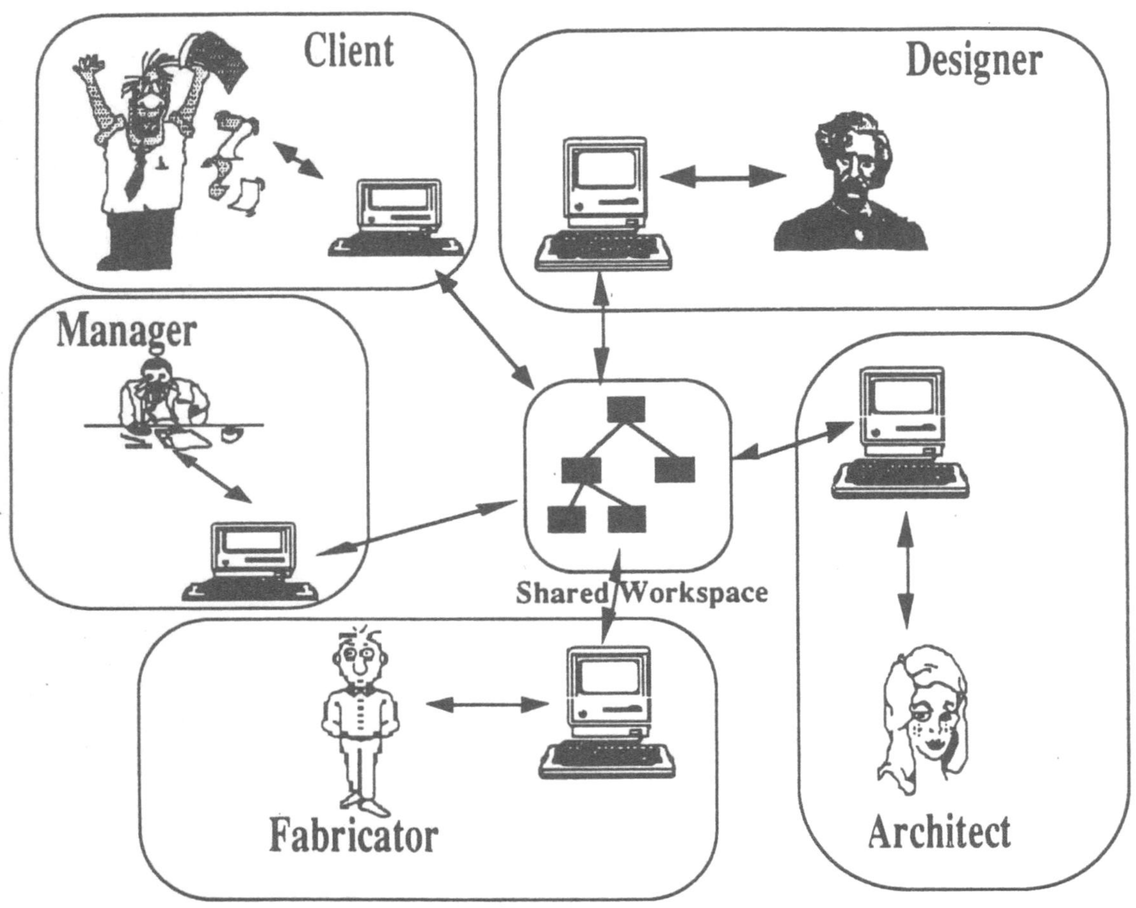

Figure 4: Computer-based View of Cooperative Product Development

1. **Frameworks**, which deal with problem solving architectures.

2. **Representation Issues**, which deal with the development of product models needed for communicating information across disciplines.

3. **Organizational Issues**, which investigate strategies for organizing engineering activities for effective utilization of computer-aided tools.

4. **Negotiation/Constraint Management Techniques**, which deal with conflict detection and resolution between various agents.

5. **Transaction Management Issues**, which deal with the interaction issues between the agents and the central communication medium.

6. **Design Methods**, which deal with techniques utilized by individual agents.

7. **Visualization Techniques**, which include user interfaces and physical modeling techniques.

8. **Design Rationale Records**, which keep track of the justifications generated during design (or other engineering activities).

9. **Interfaces between Agents**, which support information transfer between various agents.

10. **Communication Protocols**, which facilitate the movement of objects between various applications.

In the following sections we will describe our efforts in addressing some of the above problems. We will also make an attempt to compare our work with similar work in other research institutions.

3 Frameworks: The DICE Architecture

3.1 Overview

To achieve the goals outlined in Section 1, a system architecture – DICE – based on current trends in programming methodologies, geometric modeling, object-oriented databases, and knowledge based systems was developed. DICE can be envisioned as a network of computers and users, where the communication and coordination is achieved through a global database

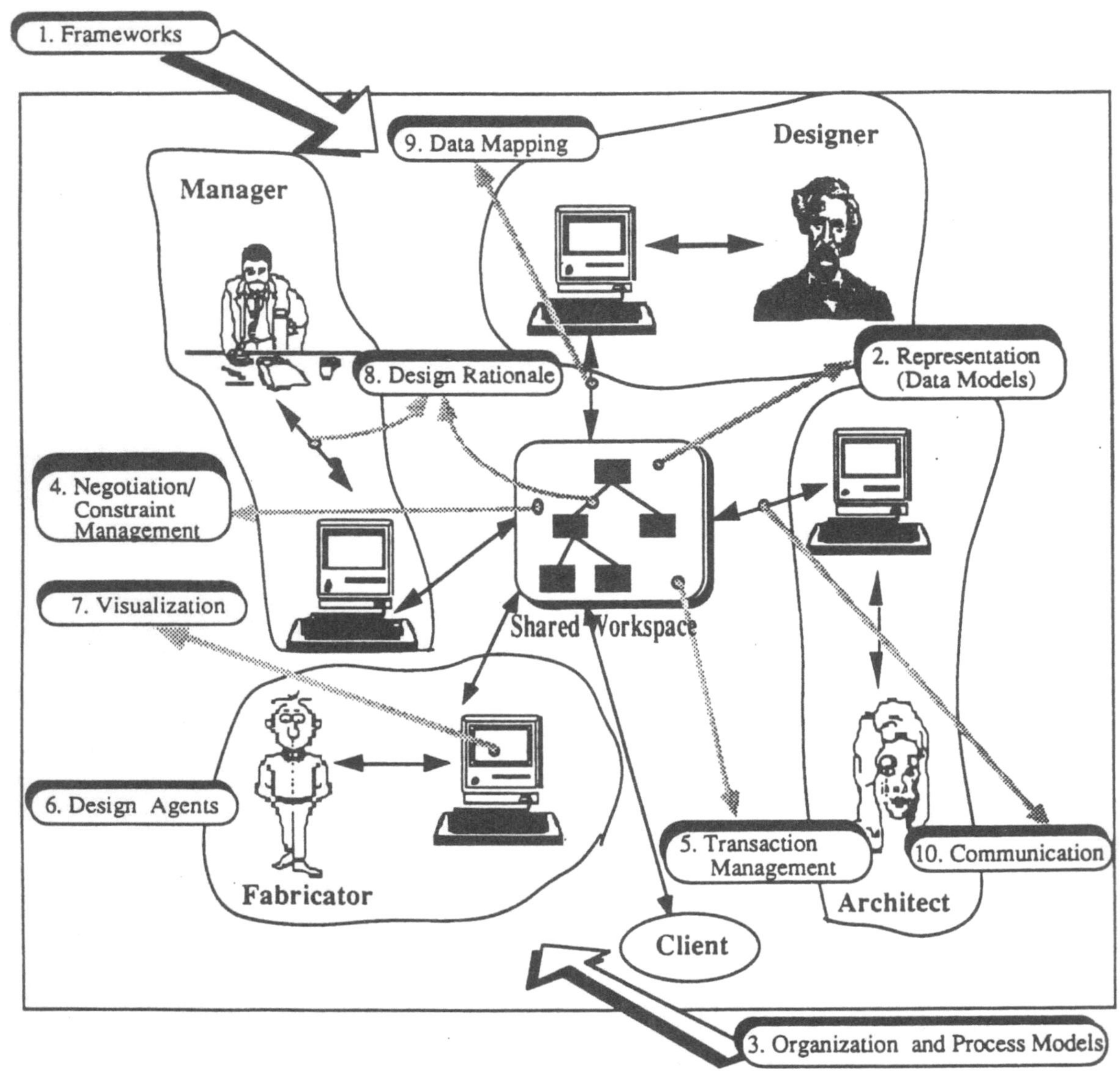

Figure 5: Research Issues for Computer-aided Collaborative Engineering

and a control mechanism. DICE consists of a Blackboard (shared workspace), several Knowledge Modules, and a Control Mechanism. These terms are clarified below.

1. Blackboard. The Blackboard is the medium through which all communication takes place. The Blackboard in DICE is divided into three partitions: Solution (SBB), Negotiation (NBB), and Coordination (CBB) Blackboards (see Figure 6). The Solution Blackboard partition contains the design and construction information generated by various Knowledge Modules; this solution is normally is referred to as the Object-Hierarchy. The Negotiation Blackboard partition consists of the negotiation trace between various engineers taking part in the design and manufacturing (construction) process. The Coordination Blackboard partition contains the information needed for the coordination of various Knowledge Modules. Note that the Blackboard could be distributed across several computers; currently we are not addressing this issue.

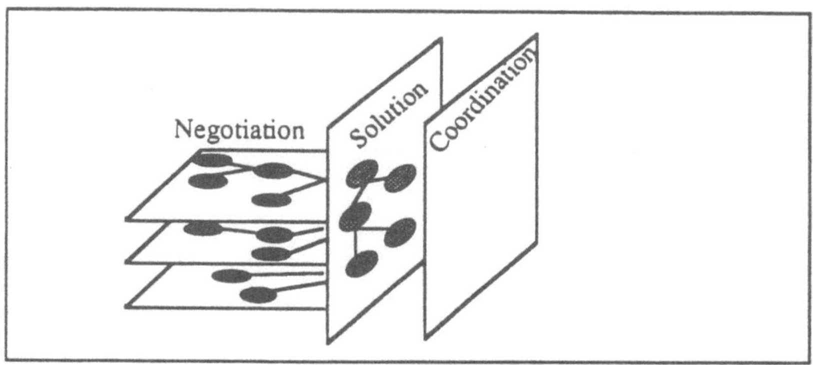

Figure 6: Blackboard Partitions

2. Knowledge Module. Each Knowledge Module (KM) can be viewed either as: a knowledge based expert system (KBES), developed for solving individual design and construction related tasks, or a CAD tool, such as a database structure, i.e., a specific database, an analysis program, etc., or an user of a computer, or a combination of the above. In DICE, the Knowledge Modules are grouped into four categories: Strategy, Specialist, Critic, and Quantitative. The Strategy KMs help the Control Mechanism in the coordination and communication process. The Specialist KMs perform individual specialized tasks of the design and construction process. The Critic KMs check various aspects of the design process, while the Quantitative KMs are mostly algorithmic CAD tools.

3. Control Mechanism. The Control Mechanism performs two tasks: 1) evaluate and propagate implications of actions taken by a particular KM; and 2) assist in the negotiation process. This control is achieved through the object oriented nature of the Blackboard and a Strategic KM. One major and unique difference between DICE and other Blackboard systems is that DICE's Blackboard is more than a static repository of data. It is an intelligent active database, with objects responding to different types of messages. A substantial part of the Control Mechanism's functionality is distributed to and localized in these active objects. In DICE's framework, any of the KMs can make changes to or request information from the Blackboard; requests are logged with the objects, and changes to the Blackboard may initiate either of two actions: finding the implications and notifying various KMs, or entering into a negotiation process, if two or more KMs suggest conflicting changes.

A conceptual view of DICE for design and construction is shown in Figure 7. In it, any of the

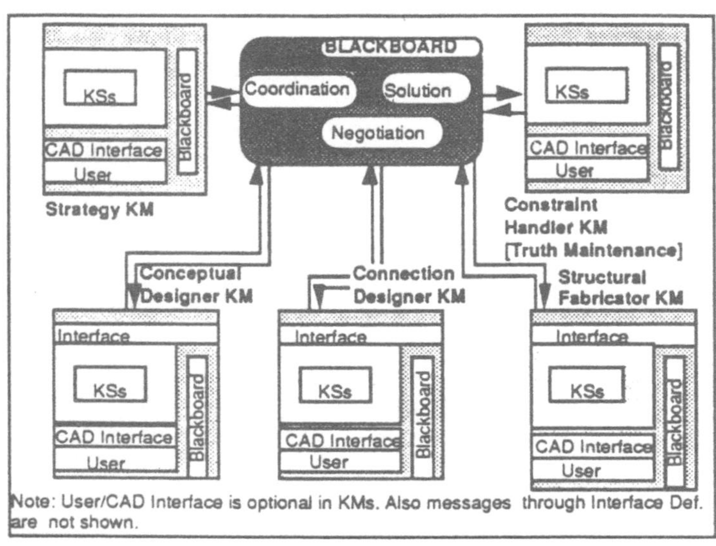

Figure 7: A Conceptual View of DICE for Design and Construction

KMs can make changes or request information from the Blackboard; requests for information are logged with the objects representing the information, and changes to the Blackboard may initiate either of the two actions: finding the implications and notifying various KMs, and entering into a negotiation process, if two or more KMs suggest conflicting changes.

An organizational view of the DICE architecture is shown in Figure 8. This view is based on the work of Moses, reported at a 1987 Xerox-MIT workshop on Visions of Design Practices for the Future.

3.2   Blackboard: Object-Oriented Database

The Blackboard is being implemented as a layered object-oriented database, as shown in Figure 9; a detailed discussion on the relevance of object-oriented database mangement

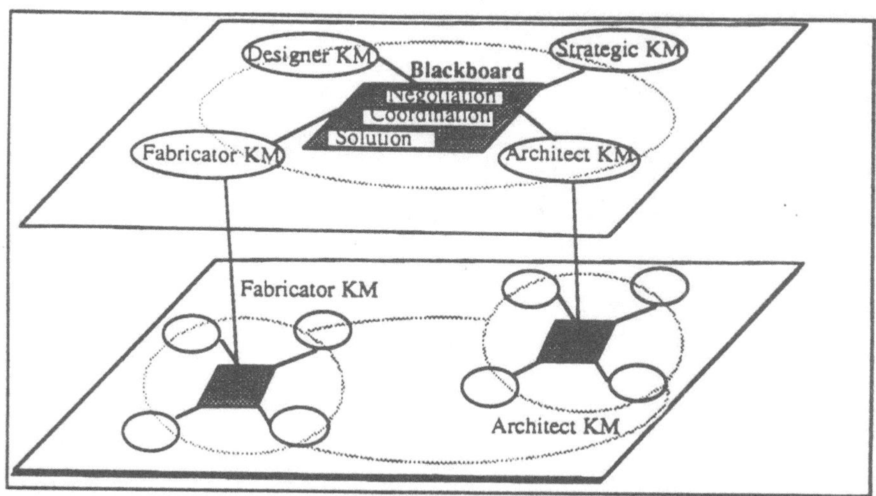

Figure 8: Organizational View of DICE
(Solid lines indicate formal modes of communication)
(Dotted lines indicate informal modes of communication)

systems (OODBMS) for engineering applications is provided in [2]. The various layers are described briefly below.

1. **Physical Layer**. Data resides in the form of bits on an appropriate storage medium (e.g., magnetic, optical, video disks).

2. **Storage Layer**. Objects are assigned physical identifiers (PIDs), which are mapped into appropriate areas in the Physical Layer.

3. **Controller Layer**. Grouping of objects, allocation and de-allocation of object buffers, and other storage control activities are achieved at this layer.

4. **Object-base Layer**. Object definition, modification, and other associated activities are included here. The semantics of various nodes and relationships needed for concurrent engineering are described in the following section.

5. **Version Layer**. Versions of objects help to keep track of the design evolution and also enhances parallelism of design activities. Various version management facilities are encoded at this layer.

6. **Transaction Layer**. Transaction management layer is responsible for maintaining database integrity, while allowing execution of multiple concurrent transactions by various engineers. This layer supports a transaction framework for concurrent engineering applications.

7. **Query Layer**. Query optimization is performed in this layer.

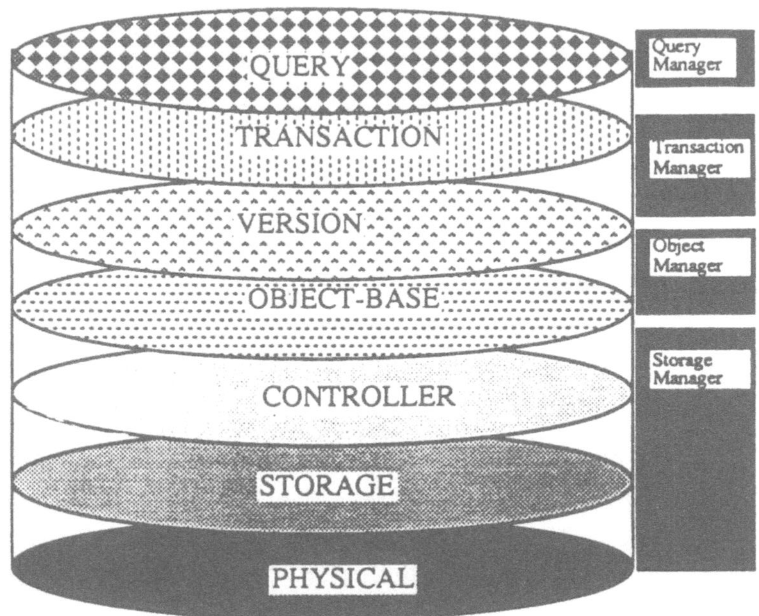

Figure 9: Layered Architecture for DICE BB

The object-base, shown in Figure 9, is divided into levels, representing an object-hierarchy (or object-lattice). Each level contains objects that represent certain aspects of the engineering process (design and construction). The SBB does not contain all the information generated by all KMs; only information that is 1) required by more than one KM, and 2) useful in the engineering process is posted on the SBB. For example, the 3D space level will contain objects that represent spaces allocated to structural systems, piping systems, mechanical systems, etc. This level can be reduced to detailed levels, such as system and component levels.

The objects in SBB are connected through relational links; these relationships provide a framework to view the object from different perspectives. Some of the relational links provide means for objects to inherit information from other objects. Representative relationships used in the SBB are (see also Section 5): *generalization (IS-A)* for grouping classes into super classes, *classification (INSTANCE-OF)* for defining individual elements of a class, *aggregation (PART-OF, COMPONENT)* for combining components, *alternation (IS-ALT)* for selecting between alternative concepts, *versionization (VERSION-OF)* for representing various versions of an object, and *association* for representing other relationships not outlined above. The semantics of these relationships are provided in [43]. Various planes that depict these relationships are shown in Figure 10.

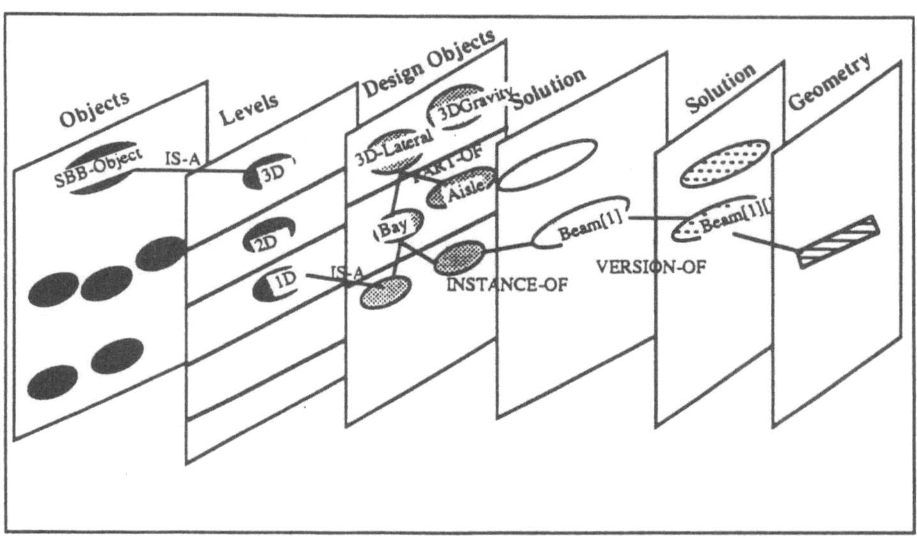

Figure 10: Different Planes in the SBB

The objects also contain justifications, assumptions,creator, time of creation, pointers to multi-media documents, constraints, ownership KM, other concerned KMs, etc. The justification information will provide a designer's rationale and intent for the creation of the object. Assumptions made during design and construction are also stored with the object. For example, an Architect, while placing the structural elements, may assume certain spatial characteristics for the HVAC systems. He may record this assumption and the rationale for such an assumption in the objects denoting the appropriate structural elements and the HVAC system. In DICE, *status* facets are associated with data attributes (slots). The *status* facet, for example, can take the following values: *unknown, assumed* and *calculated.* Additional slots needed for the source of data and its change, uses of data, assumptions made, etc., can easily be incorporated.

Associated with these objects are methods which provide a means for: 1) performing some procedural calculations; 2) propagating implications of performing some actions, for example if the status (assumed or actual) or the value for a particular object changes then these changes can be broadcast to all concerned KMs; 3) helping to perform the coordination process. We will discuss the representation issues needed for sharing information in Section 5.

3.3   Language for Supporting DICE Agents: COSMOS

An important landmark in the evolution of programming languages is the development of C++. C++ offers the advantages of object-oriented programming, while retaining the efficiency of C. However, C++ is a statically typed language and does not support the incremental addition of classes. Further, C++ does not come with a problem solving mechanism. Our object-oriented KBST - called COSMOS (C++ Object-oriented System Made fOr expert System development) - was developed to address these deficiences. In particular, our objectives for implementing COSMOS are to: 1) extend C++ to support object evolution; 2) provide persistent object store; 3) develop friendly user interfaces for entering C++ objects and rules, browsing C++ objects, etc.; 4) provide problem solving support for design agents; 5) make source code available so that parts of COSMOS can be integrated into engineering software; 6) support links to external programs; and 7) run on any Unix workstation supporting X Window/Motif toolkits

COSMOS consists of the following modules: 1) User Interface; 2) Object Manager; 3) Rulebase/Parser; and 4) Inference Mechanism. These modules are briefly described below.

User Interface. The User Interface module consists of the Expert System Development Tool (EDST) and the Expert System End User Tool (ESEUT). EDST is used by a knowledge engineer to input objects and rules. ESEUT is used by an end-user to run the knowledge-based expert system (KBES).

Object Manager. The Object Manager module is responsible for the maintenance of all classes and objects created at runtime, record keeping on the extension (all the instances) and intention (contents) of classes, access, retrieval and interaction functions at runtime on request from the user-interface and the inference engine, and persistence management of data and inference states across sessions.

Rule-base/Parser. The input to the Parser is the code generated (knowledge base) by the knowledge editor of ESDT. As its output, the Parser generates two data structures used by the Inference Mechanism. The first data structure is an inference network that is used by the backward chaining (BC) mechanism. The second data structure is an intermediate data structure, used by the RETE network building algorithm of the forward chaining (FC) mechanism of the inference engine of COSMOS to generate the RETE network.

Inference Mechanism. The Inference Mechanism consists of two problem solving strategies: forward chaining and backward chaining. The forward chaining strategy consists of a modified object-oriented RETE network.

Details of COSMOS are presented in [48].

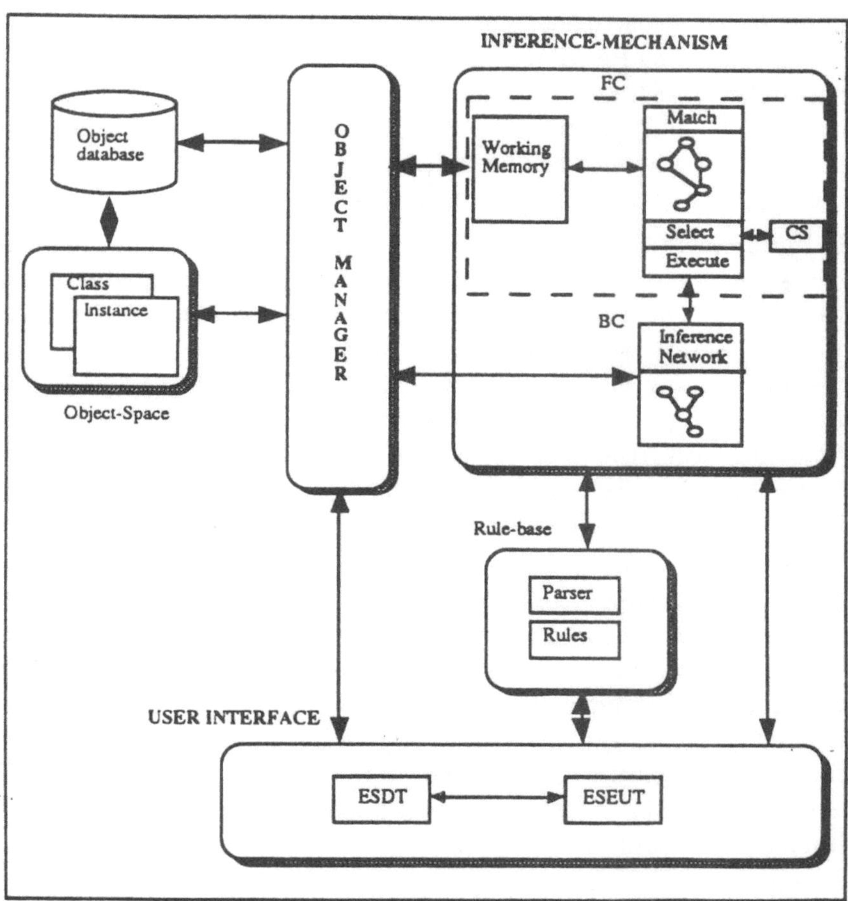

Figure 11: Structure of COSMOS

3.4   Related Work

Representative projects at other research institutions are briefly described below.

Stanford University, Mechanical Engineering. The Palo Alto Research Testbed (PACT) is an effort aimed at the development of a computational framework for concurrent engineering [12]. It is a collaborative venture between Stanford University's Mechanical Engineering and Computer Science departments, Lockheed Palo Alto Research Labs, and Enterprise Integration Technologies. The PACT architecture consists of agents that communicate through *facilitator* modules. Facilitator modules translate agent-specific knowledge into an interchange format (KIF) and communicate through a knowledge manipulation language called KQML. The PACT framework has been demonstrated on a prototype problem, which involved the design of a planar robot manipulator with several intercating design agents.

Stanford University, Civil Engineering. KADBASE was developed to provide a knowledge-based interface for communication between multiple knowledge-based expert systems and databases [28]. The main components of KADBASE are: 1) The Knowledge-based System Interface (KBSI), which provides the translations (semantic and syntactic) for each KBES for communicating with the Network Data Access Manager (NDAM); 2) Knowledge-Based Database Interface (KBDI), which provides the translations needed for each DBMS for communicating with NDAM; and 3) NADM, which decomposes queries and updates and sends them to the appropriate KBES/DB. The main emphasis in the KADBASE project has been the data translations from local to global (shared) databases/application programs, and vice versa.

West Virginia University. The DARPA DICE project uses the Blackboard approach to achieve communication and coordination. Once the agents agree on a particular design, the design is posted onto a database, which is developed over the ROSE database management system and resides on the computer network. In PACT, the differences in the representation between individual agents and the central knowledge store were not addressed in detail. The interface incompatibilities are addressed in the DARPA DICE project through the use of Wrappers, which provide appropriate translations. In the initial versions, the Common LISP environment was used to develop the DARPA DICE (see paper in [45]).

Carnegie Mellon University. Fenves et al. have developed an integrated environment - called IBDE (Integrated Building Design Environment) - of processes and information flows for the vertical integration of architectural design, structural design and analysis and construction planning [18]. The integrated environment makes use of a number of AI techniques. The processes are implemented as KBES. A Blackboard architecture is used to coordinate communication between processes. The global information shared among the processes is hierarchically organized in an object-oriented programming language.

The Integrated Building Design Environment (IBDE) system is implemented in the form of several vertically integrated Knowledge-Based processes (or design agents): ARCHPLAN,

HI-RISE, SPEX, FOOTER, PLANEX, etc.. The processes communicate with each other in two ways:

1. a message Blackboard is used to communicate project status information such as whether a process is ready to execute, has successfully performed its task or has encountered a failure, and

2. a project database used for storing the information generated and used by the processes

A controller uses the information posted on the Blackboard to initiate the execution of individual processes. The controller also directs the data manager to provide and receive the information shared between the processes. Since the different processes may reside on different machines, the data manager and the Blackboard rely on a local area communication network. The controller in IBDE is fairly domain specific; it was tool specific and was geared toward the AEC industry. IFDF is a domain independent framework that supports persistent objects through a relational database interface and provides a domain independent invocation mechanism for tools and agents, i.e., it is problem centered. IBDE-2 was developed using the facilities provided by the IFDF environment [17].

University of Illinois at Urbana-Champaign, Mechanical Engineering. The Knowledge-Based Engineering Systems Laboratory (KBESRL) has been actively pursuing research in the development of knowledge-based frameworks for concurrent engineering [35]. Their research focus – called SWIFT – is similar to ours, with primary emphasis on integration tools (based on the Blackboard approach), constraint management, negotiation framework, machine learning, and design agents. Several design agents in the field of mechanical engineering have been developed. The integration of these design agents is an on-going research project.

Industry. GE and Xerox have been working on various architectures. GE is closely tied with the DARPA initiative. The COLAB project was conceived at Xerox Palo-Alto Research Center for computer assisted collaborative work. This technology was later tried for engineering design at the Xerox Design Research Institute in Rochester. In the recent past, Xerox has teamed up with the computer science department at Cornell University to explore computer-aided collaborative design. Most of the work is at a preliminary stage and is yet to be published.

Comparison. In the DARPA DICE framework, the database and the Blackboard reside on different computers. This may cause a lot of traffic in the DICE communication channel and thus slow down the system. Our project addresses this issue by implementing the Blackboard over an object-oriented database management system; thus the Blackboard and the object-store are tightly integrated. In addition, the objects in our Blackboard have behavior associated with them; the PACT approach uses intelligent objects, but it does not seem to support persistent objects and there is no notion of a centralized data store. Hence, the need for a sophisticated scheduler (as provided in the DARPA DICE project and the IBDE environment) is obviated. Our DICE project also incorporates comprehensive transaction and version management mechanisms; other projects are only beginning to address these issues. The KADBASE project does not address coordination aspects of control. We have not concentrated on the translation mechanisms from the local to the global data models, whereas this has been the primary focus of KADBASE. The negotiation, the design rationale capture, and the conflict resolution frameworks in the KBESRL project are superior to ours. However, our OODBMS-based Blackboard has several advantages over the KBESRL approach (e.g., persistent objects, transaction management, active objects, support for multi-media, etc.). Although, in the recent past the SWIFT project has shifted to a commercial OODBMS as a backend. Another important difference is the language of implementation. We are developing our DICE tools in C/C++ and Motif/X Windows, which makes our system very efficient and portable; LISP is the language used in many other projects, which may limit their use by the industry.

4 An Example Of Collaborative Engineering In The DICE Framework

We will illustrate our framework with a simple example, that of collaborative development in building design and construction.

4.1 Design Agents

The design of a building involves teams of several designers, different technologies, and components. There is considerable necessity for controlled interaction and cooperation between different design groups for the successful completion of the design task. Some of the design technologies and agents in the DICE framework are shown in Figure 12. These include (among others): project manager, architect, structural engineer, geotechnical engineer, HVAC (heating, ventilation and air-conditioning) engineer, electrical engineer, plumbing and sanitary engineer, fabricator, contractor, owner, etc. We will consider a very *simplified* scenario involved in the design of a small building.

4.2 Design Components

The various components of a building are shown in Figure 13. It consists of a superstructure and a substructure. The architect is responsible for designing the skeletal plan of the

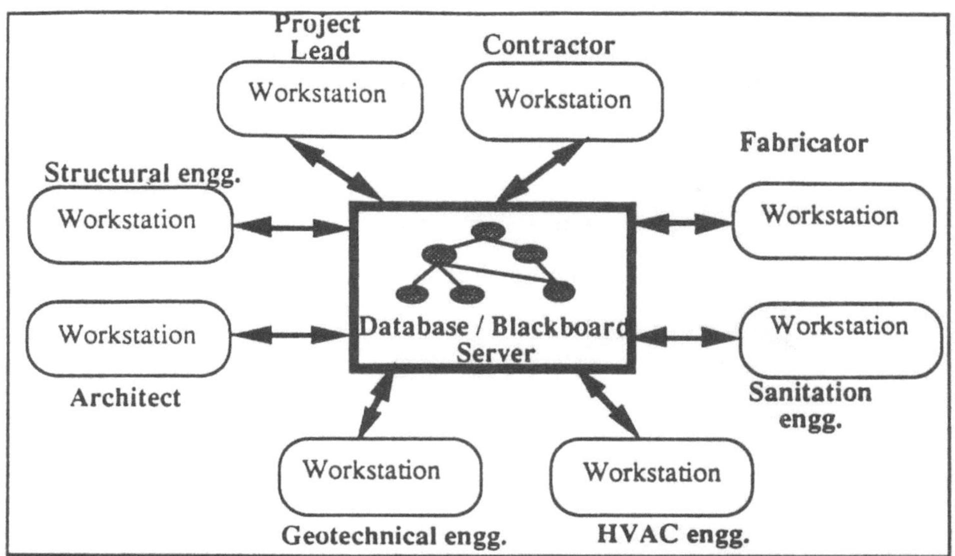

Figure 12: DICE Framework for Agents involved in Building Design

building, locations of beams and columns, layout of rooms and hallways, interior design, etc. The structural engineer takes specifications from the architect to design the superstructural elements of the building, such as beams, columns, slabs, connections, stairways, joints, etc. The geotechnical engineer takes column loads estimated by the structural engineer to design the substructure of the building. This includes components like the footing, foundation, water-proofing, and water-retaining structures, etc.

The house is modeled as a *composite object* ([4], [46]) in the DICE object-oriented database. The **House** is a *containing* object which is composed of *component objects* such as **Superstructure** and **Substructure**, which are themselves composed of several other objects such as **Beam**, **Column**, etc. (Figure 13).

4.3 Database Organization

The principal agents involved in our simplified scenario are the architect (**A**), structural (**S**) and geotechnical (**G**) engineers. As envisioned in the DICE framework, these agents work on individual client workstations on a network with the database (or blackboard with control mechanisms) residing on a server machine (Figure 12).

Since collaborative engineering entails data and information *sharing*, it necessitates *partitioning* of the database into local shared areas. The database architecture for our example

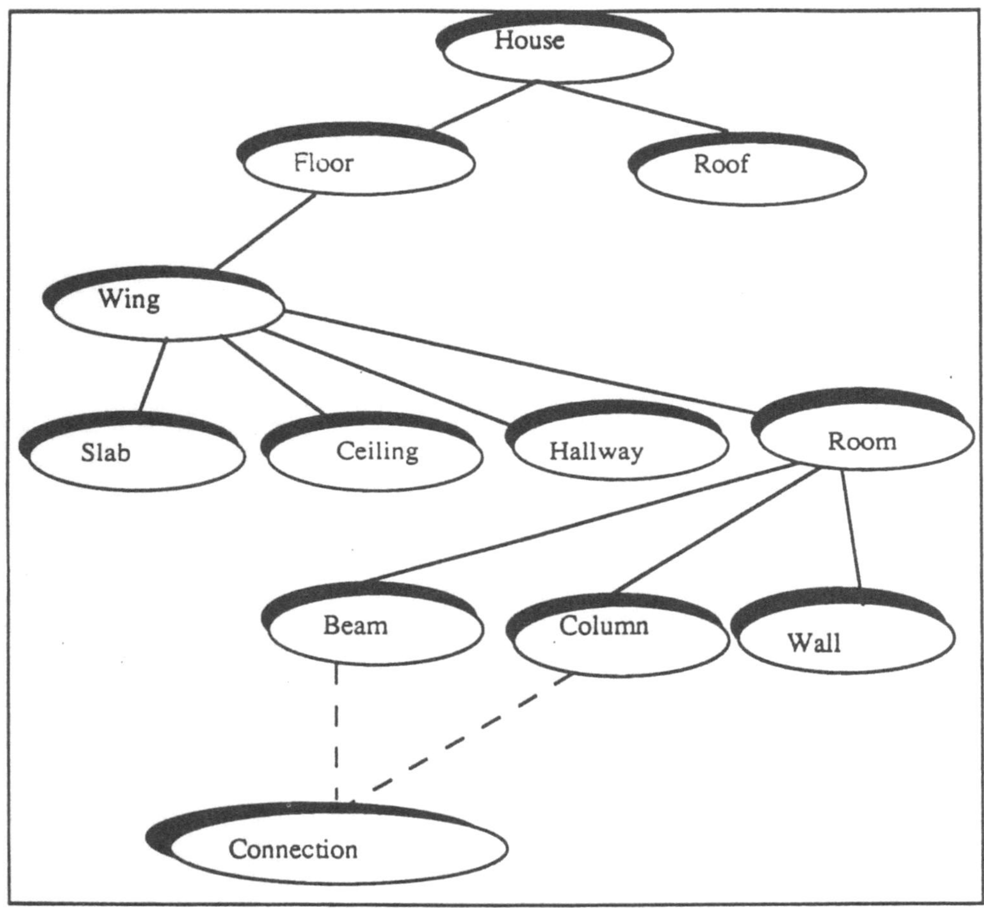

Figure 13: Building components

(at some stage of the design process) is shown in Figure 14. Each designer has his/her own *private* dataspace for doing work that is not accessible to others. When designers need to cooperate, they work on local *shared dataspaces* that are derived from the global database (or subdivisions thereof) that are read/write accessible to all of them.

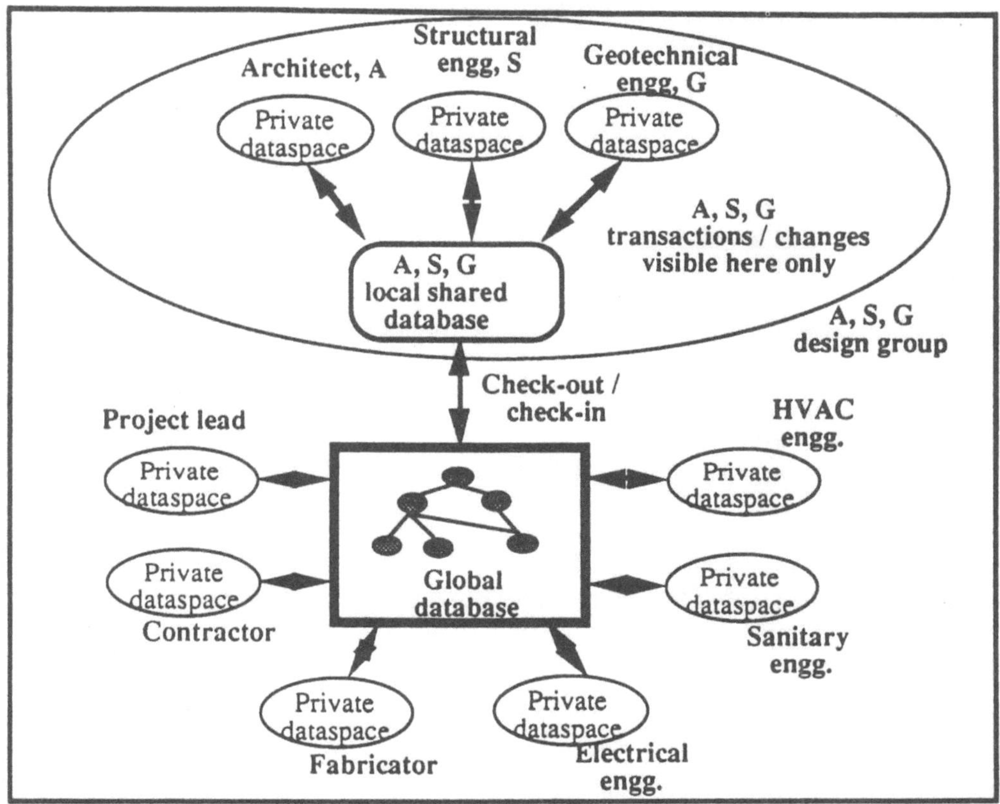

Figure 14: Illustrative Database Partitioning for Design

In this example, **A, S** and **G** are working as a *design group* and share a local dataspace which at startup, is either empty, or contains a *copy* of all the relevant objects (required for design) taken from a consistent database.

All changes made to these objects are visible only in the scope of the shared database, so intermittent changes during the group's design effort *do not affect* other designers not concerned with the group's activities. When all designers are satisfied with their design, appropriate objects in the shared database will be released to the global database (or to the parent dataspace from which the local database was derived), so that the new objects can be shared with others in the group.

4.4 Collaborative Design

The following represents the steps and interactions by which the design would normally proceed (Figure 15 a, b). It is assumed that each designer is assisted by a set of automated analysis and design tools (represented by knowledge modules in the DICE framework), such as analysis and design packages.

- The architect (**A**) designs the skeletal layout of the building, positions of columns, beams and preliminary sizing of these components. This information can get the structural engineer started with his/her design, so **A** posts these results to the database. **A** may include dimensional or other *constraints* on the various building component parameters that **S** may have to abide by. In the meantime, the architect (**A**) can continue with details of internal layout, such as walls, partitions, etc.

- The structural engineer (**S**) is notified of the posting, retrieves the **House** from the database and proceeds with the preliminary design, such as estimation of live and dead loads, sizing of the components, etc. At this stage, an estimate of the loads on the columns is known, so the geotechnical engineer can start with preliminary substructure design. **S** therefore posts his preliminary design to the shared database. **S** then continues with detailed structural analysis and design.

- The geotechnical engineer (**G**) is notified of the posting. S/he retrieves the *appropriate components* (e.g., the column objects) from the database and proceeds with preliminary foundation design, such as the distribution area required for pressure dissipation, type of foundation required, etc.

- As **S** proceeds with detailed top-down structural design, the column loads become better known. After the design of each floor (from top down), **S** posts results to the database, which gives **G** notifications of better estimates of loads on the foundation. **G** may then proceed with the detailed design accordingly or refine previous preliminary designs. By the time **S** is finished with the detailed design of the superstructure, **G** will have finished a considerable amount of work on the sub-structure as well.

- As **A** proceeds with detailed layout, s/he may find it necessary to make modifications to previous designs. **A** posts these changes to the database, so that **S** is notified immediately. **S** can immediately look into the changes made made by **A** to check whether it necessitates changes in his/her design. If so, changes can be effected, and the implications passed on to **G**.

  It may be noted that whenever changes are made to existing data (or objects) in the database, the old data is not overwritten, but a new *version* of the object is created (provided the object is versionable). This preserves the *design history* and enables the design to be restarted from a given specified state.

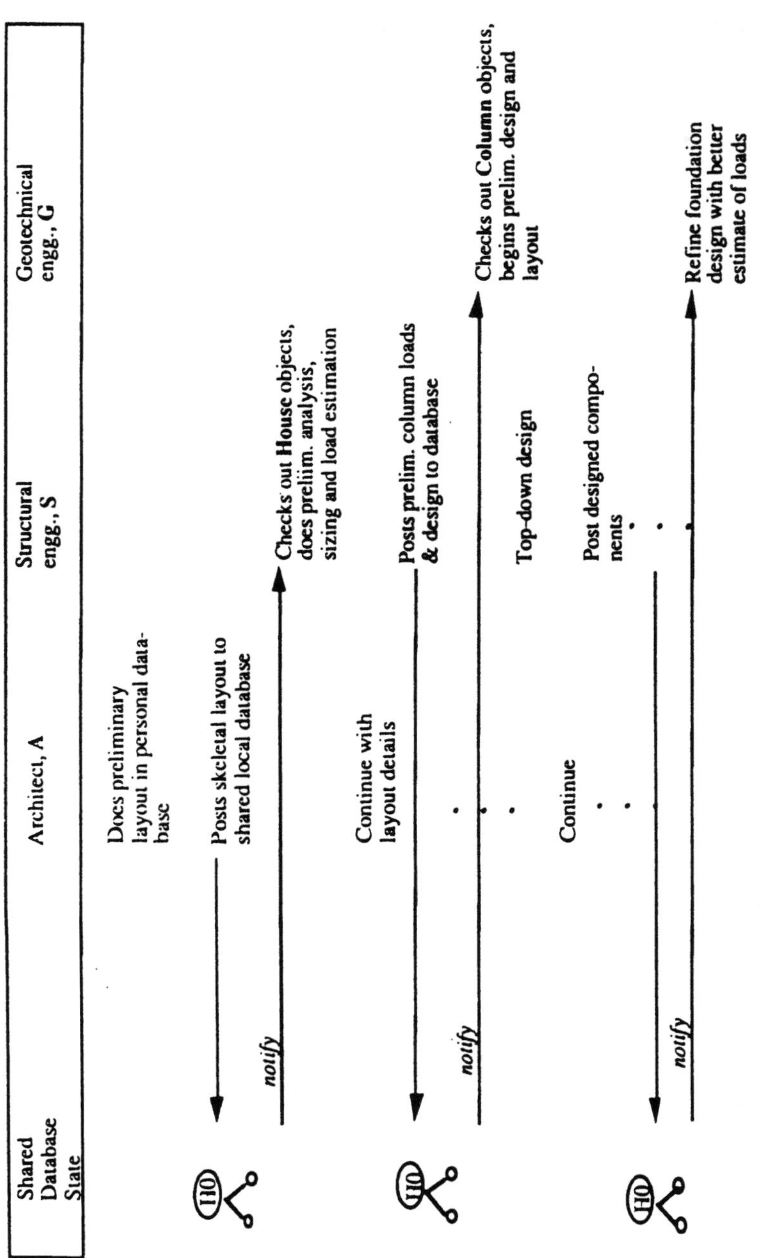

Figure 15: Design Steps in Building Design (simplified)

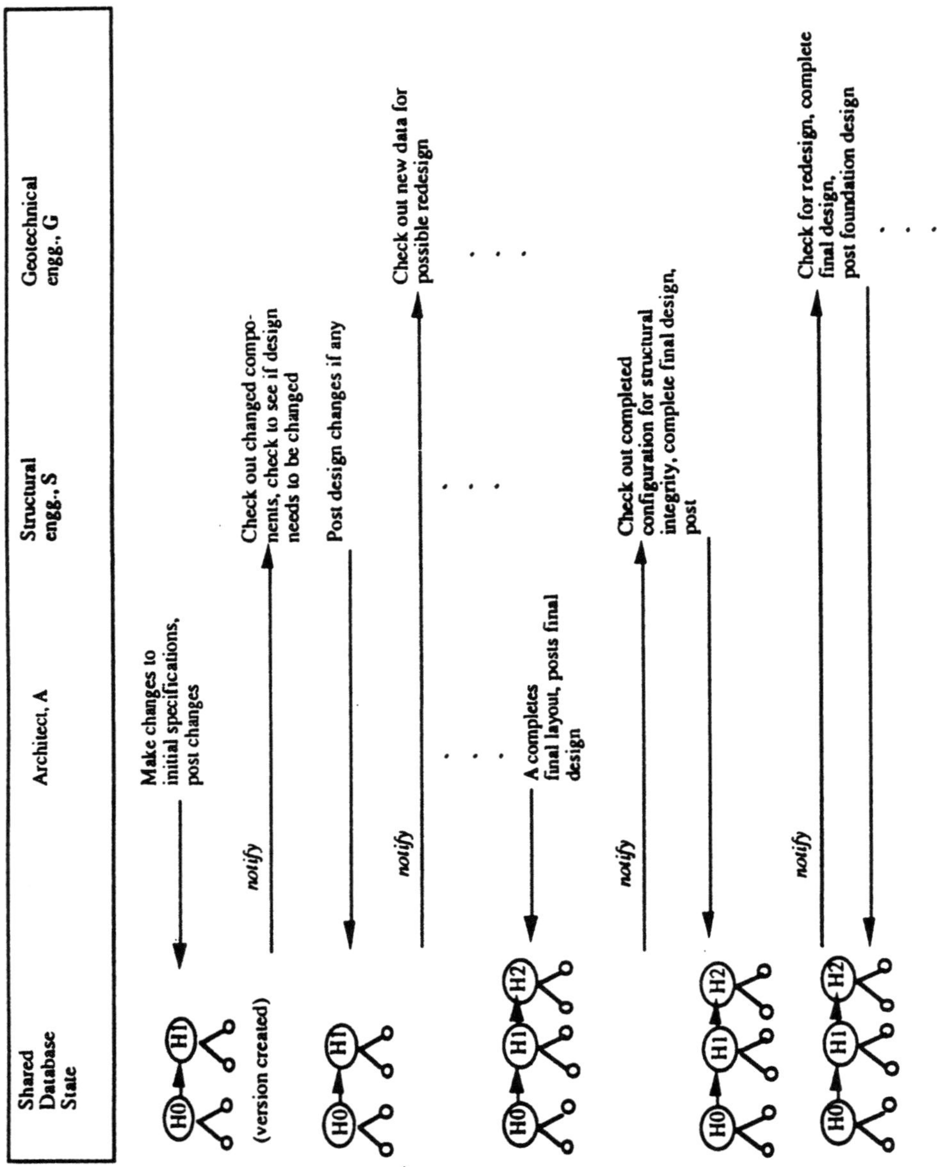

Figure 15: Design steps in Building Design (Continued)

- When the architectural layout is complete, **A** posts the entire layout to the database. This configuration is retrieved by **S** to check for structural integrity of the superstructure. If it passes the check, any changes in column loads are passed on to **G** who refines the substructure design.

- In the event that there are anomalies or *conflict* between designers, there is a necessity for closer scrutiny of the design, greater degree of interaction, and possibly *negotiation* for an agreeable design. For example, **A** and **S** do not agree on the dimensions of a beam, or the design requires dimensions that are beyond the range specified by **A**. In such cases, **A** and **S** may form another *nested design group* between themselves with a smaller shared dataspace containing only the relevant objects, and resolve the conflict by negotiation. If active experimental interaction is required, **A** and **S** may participate in a *shared transaction*. The concept of a shared transaction will be explained in a later section; it suffices here to say that **A** and **S** may initiate a common transaction between themselves rather than having to communicate *across* transaction boundaries.

- When all design components have been agreed upon, the appropriate objects are checked out into the global database, where a new version of these design objects is created. These new designs may then be shared with other members, such as HVAC and electrical engineers. Typically, the architectural, structural and HVAC engineers would then form another design group to complete the HVAC design of the building, and so forth.

4.5   Key Features

The following points of interest may be noted regarding the methodology described above:

1. It may be noticed that the above scheme allows engineers in different disciplines to proceed with their work in parallel, although the preceding design group has not completed its task. This is because the amount of information that is necessary to get other engineers in the group started is released early, and is *communicated* to those concerned. This significantly increases the concurrency of design effort.

2. The system maintains records of data used by various designers and establishes *dependencies* between data and clients. Thus, changes to data or objects result in notification to the appropriate designers who have used or accessed the data before. For example, a change in the **Column** object by **S** would immediately notify **G**. Thus, the designer *does not have to look out for changes*, his/her attention is drawn automatically when appropriate pieces of data are changed.

3. With several designers collaborating in design effort, and changing data interactively, it is necessary to embed *validity constraints* on design data in the database. Whenever these constraints are violated by an update, the designer(s) concerned are warned of the violation, so that they may backtrack and redo the design appropriately. Constraint violations also need to be logged, and reported to those who have set them. For example, **A** may set a constraint such that the perimeter of a column should not exceed **n** units. When **S** posts a column design, this constraint is checked for validity.

The above framework provides a versatile and flexible platform to enable and coordinate collaborative design in most CAE disciplines. The following sections describe our work on shared workspaces, transaction management, user interfaces, and design agents.

5 Representation: The Shared Workspace

One of the basic issues in developing collaborative engineering systems is the representation of the product information which supports sharing (the product model resides at the Object-base level). This product information includes not only the geometric data of the physical parts of the product and their relationships but also non–geometric information such as details on functionalities of the parts, constraints, and design intent. Requirements for such a design representation are: These include:

- **Support for multiple levels of abstraction and different functional views.** This is needed to allow a top-down design process which involves the refinement of levels of functional abstraction into physical parts;

- **Support of multiple levels of geometric representation.** Geometric and topological information are an important part of design. However, at different stages in the design process, different levels of geometric representations might be required; and

- **Management of constraints.** Constraints between the different representations and abstractions during evolution of the design should be properly managed. Constraint management facilities could help in maintaining the integrity and consistency of the database.

Besides, the representation should be reasonably general, hence allowing for the addition of new abstractions or physical components without requiring extensive changes. Furthermore, there is the requirement for mapping it to a distributed database environment which supports persistency and concurrent access in the shared workspace.

Our work aims at providing a framework for representing product information in a shared workspace which supports the requirements outlined above. The focus is on the development of general concepts such as geometric representations and abstractions of general properties such as concepts of "compositional" hierarchies of systems and components. These can be seen as primitives in our model, which is called SHARED. SHARED is implemented over a commercial OODBMS and utilizes GNOMES, which is based on a non-manifold boundary representation - the *Selective Geometry Complexes* (SGC) model [40], and COSMOS.

5.1   The Sharable Primitives of Our Representation

The various primitives defined in the representation are:

The Composition relationship. The composition relationship defines a special relationship between a composite object and lower-level objects which together make up the description of the composite object. For example, a house can be "composed–of" a number of floors which in turn is "composed–of" of different rooms. Special semantics are associated with this relationship, such as the definition of interface attributes and default attributes which are to be associated with the component objects, the dependence of component objects on the existence of the composite object, constraints between the description of the composite object and its components, and expressions for (accumulating) deriving composite attributes from its components, etc.

System.   **System** is the base class from which the higher level functional abstractions are specialized from. Systems are "composed–of" subsystems (systems) and components. Systems also have an associated attribute – **Space** – besides other inherited attributes. The constraints include restriction of the sub-spaces descriptions to be contained in the description of **Space**. **System**, and all other primitive classes, also contain a set of access and constraint methods.

Space.   **Space** is a class which encapsulates the different levels of geometric abstraction representing a system. These levels of abstraction are solids, 2D-sections (plan, elevations, and other sections), lines, and symbols. It defines methods for navigating between different spaces, area, volume, intersection calculations, display, etc.

Component. **Component** is a class representing the functional abstraction of a physical object. It can be considered as a wrapper around the physical objects, providing additional domain specific information or constraints.

Physical Object.   The phsyical objects are objects in the lowest level in a functional abstraction hierarchy and represent actual physical objects which cannot be broken down into smaller parts in standard construction practice. The main attribute is an instance of the **Physical-Description** class.

Physical-Description.   **Physical-Description** class is inherited from the **Space** class. It provides mainly physical descriptions such as geometrical and topological details, spatial relationships between parts, material type, and finishes. Together with the physical objects, they have behaviors (methods) which allow processing of physical properties (e.g., calculation of weights and maintaining constraints of geometry) and methods for maintaining functionally independent geometric and topological relationship (constraints) between different physical objects or between different geometric abstraction, and for schematic displays. The physical objects are also identified by standard trade indexes, such as those of CSI's MASTERFORMAT. To provide for generality, the physical properties of the physical objects are modelled with an augmented non-manifold geometric model [54].

The classes described above provide a common foundation for the information sharing between the different functional domains. This design information can be considered to be separated into different *Functional Spaces* (See Figure 16) containing the abstraction hierarchies of each domain with classes specialized from **System** and **Component**, and a *Physical Space* containing physical objects which are linked to the components. These spaces can be mapped directly to distributed databases. Figure 17 shows the definition of an Architect's **A_Floor_ceiling_sys** system (afcs1) in a particular room of a floor in a typical bay (for further details see [53]).

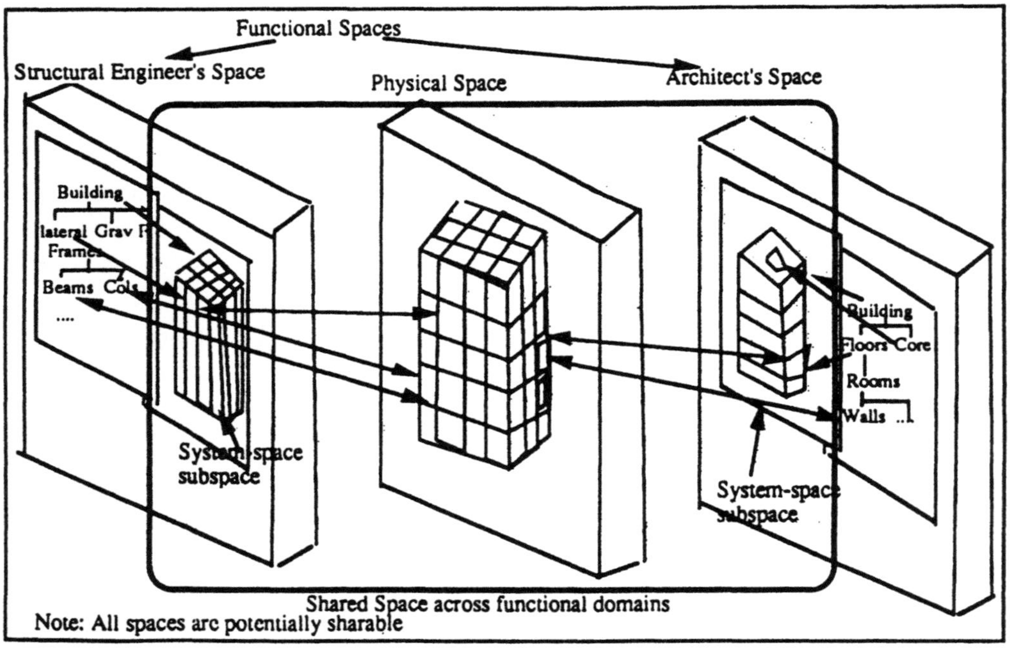

Figure 16: Conceptual organization of the Product Model

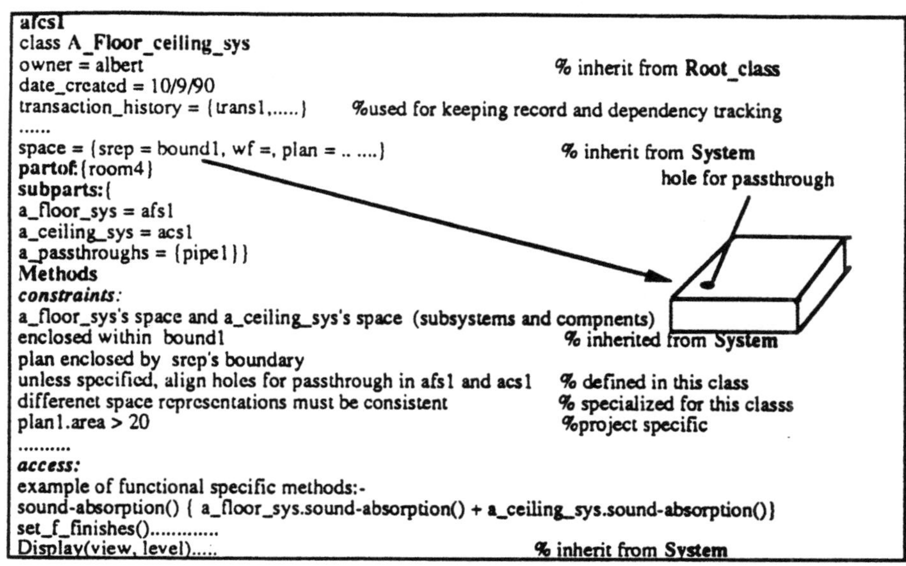

Figure 17: Definition of an Architect's Floor-ceiling System

5.2   Related Work

The STEP/PDES effort, the RATAS project [7], the EDM model [16], and the spatial representation work being pursued at Carnegie Mellon University [55] are relevant to our work.

STEP/PDES. STEP/PDES is an international initiative aimed at standardizing intelligent CAD data, where design objects are interpretable by computers (not only human). The CAD data also includes data other than geometry. A formal language called EXPRESS has been defined in STEP for encoding the design data. The relevant work in this effort for AEC is GARM (General AEC Reference Model) [23]; the GARM model is yet to be accepted by the PDES/STEP AEC committee. GARM is a very abstract high–level conceptual model which provides constructs for the modeling the complete life cycle of most products. The basic entity in GRAM is a Product Definition Unit which has various characteristics (attributes) related to an aspect (something of interest functionally, i.e., cost, strength, safety). Relationships that are defined are specialization (which specializes PDU into more meaningful entities for specific product models), decomposition, classification, and occurrences (instances).

Our emphasis has been on the modeling, and maintaining of topological (spatial) and geometric constraints between different representations of an object (as it evolves), in the same level (e.g., spatial relationship between physical objects), and between different levels in the composition hierarchy. Another difference is the emphasis on functional access in a shared database while STEP seems to be at the moment more inclined toward sharing data through standard files format. GARM's concepts of Functional units and Technical solution map the design process more explicitly. However, note that in our scheme, the Functional units are defined in the system definition through a the functional attribute. For example, a lateral system's function is to resist lateral load and the actual lateral system instance with its space representation is a technical solution which must satisfy the specification set in the building instance. The various specializations of a lateral system (e.g., frame, tube, shear wall) provide the several alternatives, provided they satisfy the constraints (or the assumptions) specified in its composite system.

RATAS. RATAS model is similar to our conceptual model with the following levels: Building (one per building), Systems (only one level in a building), Subsystems (can be multiple levels, subsystems can be composed of subsystems), Parts, and Details. However, not much detail is available on how these classes are modeled. Geometrical representations are not considered in RATAS.

EDM. In EDM, the basic unit is a functional entity (FE) which contains an aggregation of attributes and a set of constraints. These functional entities can be grouped into physical objects (with constraints) which represent engineering products. A physical object can also be related to a set of physical objects through the composition relationship. Constraints can be defined on this relationship. An accumulation relationship is also defined between a FE and a set of FEs, which allows constraints and expressions for determining values from the containing set of FEs to be defined. Accumulation is used to aggregate data to a higher level of abstraction. The emphasis in EDM is on the higher level organization of information while our focus is on the modeling and management of geometric properties. Besides, by not using actual physical components at the leaves of the product hierarchy, considerable complexity and redundancy might result. Implementation in current OODBMS is also less complex with our model.

Spatial Representation. Zamanian et al.'s work is more focused on the representation of spatial abstraction (geometry) of spaces (occupied by a design object) and their organization. A general spatial representation scheme is devised and implemented on top of a non-manifold geometric modeler (NOODLES)[25]. A relational database is used for non-spatial attributes. The scheme basically considers a configuration of n-dimensions to consist of a number of n-dimensional spaces and i-dimensional ($0 <= i < n$) partitions. Both spaces and partitions are geometric elements (known as superior elements); the superior elements are fairly general and do not correspond to any component and hence need not be linked to the relational database. In other words, different functional view of spaces can be identified by a set of superior elements which *may* be labeled and indexed with non-spatial attributes in the relational database; actually their implementation has an object-based layer between the geometric modeler and the relational database. An object–oriented approach (which is an extension of the object-based approach), such as ours, would be more suitable for several reasons: 1) constraints (implemented in the form of methods) between the different views can be checked or maintained; 2) a more explicit and integrated representation of functional decomposition is facilitated; 3) non-spatial attributes, such as color, finishes, etc., can be stored with the augmented topological/geometric element; and 4) more flexible transaction and version management facilities can be encoded [3].

Content Data Model. The Content Data Model (CDM), developed by the Air Force Human Resources Laboratory, is an attempt to provide a neutral database of information for various maintenance activities. The data is organized as a hierarchy of objects, which represent various kinds of information: text, video, graphics, etc. A weak support for composite objects is provided in the model. The model is still evolving and is yet to be used in a commercial environment, though several prototypes exist (one such implementation was co-supervised by the author). CDM lacks a number of features needed for product development; its main emphasis has been on the maintenance aspects of the product.

Other related work include the primitive-composite-model [27], being developed at Stanford University, resource integration in the Carnot project at MCC [10], the ESPIRIT's COMBINE project, the KIF and KQML projects at Stanford University and University of Maryland (see references in [12]), and the metamodel [31], being developed at Tokyo University. The primitive-composite-model does not address collaborative issues and is more limited in scope. The Carnot project and the metamodel work are more ambitious in scope than the SHARED approach. They focus more on encoding fundamental knowledge structures, from which a global schema can be generated automatically.

6   Transaction Management

The transaction management system is responsible for maintaining database integrity while allowing execution of multiple concurrent transactions by various clients. The primary functional modules of our transaction management system include (see Figure 18):

1. Transaction Scheduling. This is responsible for initiating, queuing, executing, logging, terminating or aborting transactions.

2. Lock Management. This allows locking of objects and classes for the purposes of reading and writing. Various locking modes include *read, write* (in restrictive and non-restrictive forms), *exclusive*, etc. It also maintains a record of the lock status of each object and the clients holding those locks. Lock requests are *queued* if they cannot be granted immediately, and notifies clients in the event of conflict, so that they may take appropriate action. A flexible locking protocol allows lock requests and releases anytime during a transaction rather than according to the *two-phase* protocol (details of the two-phase protocol can be found in any DBMS book).

3. Deadlock Management. This detects deadlock between transactions and notifies ap-

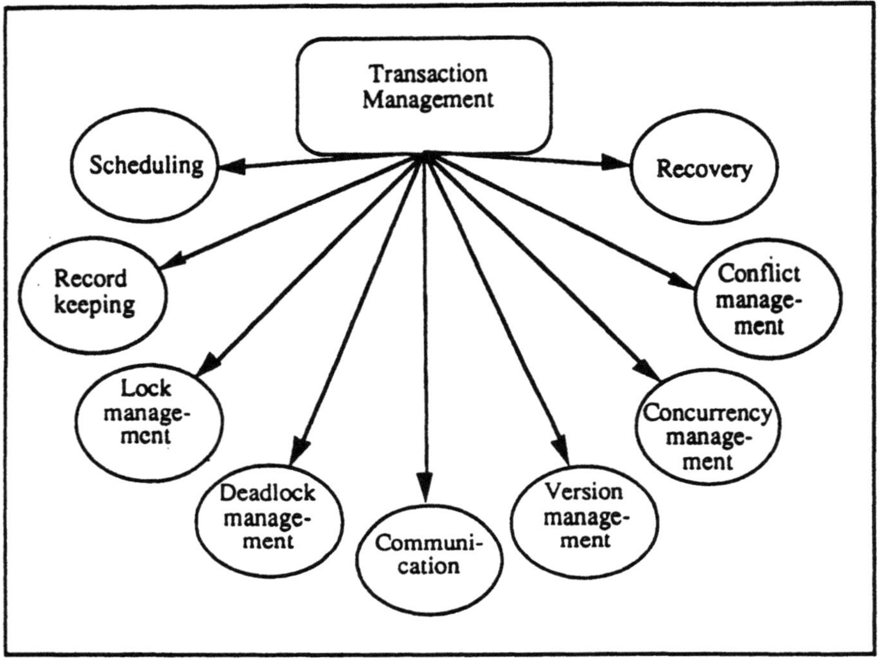

Figure 18: Functional Modules of the Transaction Management System

propriate clients so that they may communicate between themselves and resolve the problem in an agreeable manner. System-dictated transaction aborts are avoided as far as possible.

4. Communication and Update-notification Facilities. These allow communication between clients so that they are aware of each others' developments and database changes, and enable better synchronization of work across design interfaces. Various communication modes include (a) *lock modes*, where affected clients are informed in the event of lock conflicts; (b) *update modes*, where *dependent* clients (who have accessed a given object at any time) are notified of all changes to the object, the nature of the change and the id of the changer; (c) *conflict modes*, which notifies clients of the nature of the conflict, such as deadlock, transaction commit, etc., so that they may resolve the conflict; and (d)*negotiation modes*, where designers negotiate for a mutually agreeable solution (for example, trying to merge two different design versions of an object). Multimedia negotiation platforms with video conferencing, images and hypermedia text help to expand the bandwidth of communication between clients.

5. Version Management. This keeps a record of data changes and design evolution by creating versions of objects that have been updated in the course of design. It promotes greater concurrency by allowing different clients to work on their own versions of an object simultaneously, and later *merge* them together, rather than having to wait for each other to release the object. OODBMS may support versions of instances, classes and class hierarchies (database schema).

6. Conflict Management. This detects conflicts between database clients and their operations and helps to resolve them using various communication protocols. The intention is to allow the clients to resolve the conflict in a most semantically reasonable manner and avoid arbitrary application aborts by the system.

7. Concurrency Management. This allows concurrent transactions to interact and ensures database consistency according to various criteria, such as, type-semantics, operation-semantics and serializability. Serializability is too restrictive a correctness criteria for collaborative CAD operations and may be effective only for short duration transactions. Transaction *visibility* (the ability of transactions to interact) is enabled by features such as *object registration* and *checkpointing*. Object registration enables a transaction which acquires a write or exclusive lock an on object (making it inaccessible to others) to make its latest state, at any point in the transaction, visible to others *without having to commit the transaction or release the locks on the object*. It sends appropriate notification to all affected users of the object. This is particularly useful for transaction groups which are actively sharing their results with each other. It also enables a "snapshot" reader of the object to remain up to date without acquiring a restrictive lock on the object. Checkpointing a transaction at any point in the transaction commits all the changes made up to that point irrespective of whether the rest of the transaction aborts at a later time. This helps by saving changes made during long transactions instead of waiting for it to complete. These schemes clearly violate the principle of atomicity of transactions as a measure for maintaining database consistency. However, OODBMS type systems and operations may incorporate a great deal of *application semantics* which, along with intelligent programming, may help to define the notion of data "correctness" beyond the limited criteria set by a sequence of ordered read and writes [3]. Most object-oriented operations might interact in ways that are at a much higher level and quite different from traditional reads and writes.

8. Transaction Nesting and Grouping. This allows complex transactions to be divided into nested sub-transactions, which may be grouped together for active data sharing and interaction. Functional sub-division of a transaction is also accompanied by *structural partitioning* of the database into different local and global areas. Nested and grouped transactions interact between themselves and with their parent transactions through a set of *protocols* which define the rules for data access and visibility, locking, etc. for each group. Nesting enables transaction management at a higher and simpler level of abstraction. It also reduces interaction traffic at the database server, since each of the nested transactions communicate their requests through their parents at every level.

9. Recovery Management. This provides facilities for database restoration from soft and hard system crashes by persistent logging and shadow paging techniques.

6.1   Related Work

Related work in the area of collaborative engineering transactions includes models for semantic concurrency control and alternative correctness criteria nested transactions and transaction groups. Recent work that we will consider briefly are those of [5, 34, 32, 33, 42].

Korth et al. [32, 33] present a transaction model for long duration CAD applications. The model discretizes a project transaction into a multi-level hierarchy of *client/subcontractor transactions*, each of which is composed of a set of serializable short duration transactions. However, the model relaxes the requirement of serializability by replacing it with *predicate-wise two-phase locking*. This allows a schedule of transactions to be considered *predicate-wise correct* if a set of constraints (or predicates) on a set of data items is preserved, even though the transactions may not be serializable. This provides greater flexibility than normal two-phase locking. However, these predicates cannot be redefined within the scope of sub-transactions to allow greater localized concurrency that is possible within transaction groups. The model does not address *communication* between cooperative designers and their transactions as a mechanism for sharing transient design information. For example, it is not possible for a designer to read the current status of a design object that is continually being modified by someone else, since there is no notification mechanism. This lack of communication compromises collaborative effort.

Skarra [42] discusses a cooperative transaction model where application programmers define a set of semantics based *correctness criteria* and and *correct concurrent histories*. Concurrent transaction schedules are considered valid if the execution history matches one of the correct histories, even though the schedule is non-serializable. The correctness criteria may be defined locally within each transaction group, which enables enhanced localized concurrency (by relaxing the correctness criteria) inside a transaction group without affecting the global data consistency. Correctness criteria are declarative descriptions of valid histories in the forms of *patterns* and *conflicts*. A *pattern* describes invocation sequences that are *required* for correctness, while a *conflict* describes sequences that are prohibited. A transaction schedule is admissible if it contains all the invocation sequences in a pattern and none of those in a conflict. Patterns and conflicts are described in a construct called a *pattern machine* which is a finite state automation with the ability to evaluate predicates, to perform actions such as updating variables and sending messages, etc. This provides a flexible and powerful mechanism for controlling concurrency as compared to traditional serializability-based approaches. However, it requires that the "correct" sequence(s) be known aforehand and all invalid sequences identified. It is therefore useful for routine design applications and for non-serialized transaction handling in non-interactive processes. The model does not address the issues of design version management and communicative locking of data.

Barghouti and Kaiser [5] have developed a flexible transaction management framework which incorporates several levels of processing. The key feature of their system is a rule-based environment that can be programmed to deal with specific projects. The transaction management system is a part of a multi-user CASE environment – MARVEL. MARVEL also supports the notion of split-transactions, which addresses the dynamic transactions issue. We are currently evaluating the use of MARVEL for DICE.

Kutay and Eastman [34] present a model which ensures the database integrity after a sequence of transactions. An inappropriate sequence of operations by agents may lead to loss of data integrity. The entity state transition management scheme facilitates the partial ordering of transactions, which ensures database integrity and aids in concurrency control. A pipe layout design, with an emphasis on theoretical foundations, is used to exemplify the proposed scheme. Some of the concepts presented are complimentary to our work. However, it does address the issue of cooperating transactions, neither does it exploit the semantics of OODBMS.

7 Visualization: User Interfaces

7.1   Introduction

The user interface for DICE can be considered as a set of tightly integrated high level tools which provide facilities required for cooperation between various groups. It makes use of multiple windows, menus, graphics and direct interaction techniques to provide a graphical interface with an organizational structure which is easy to comprehend and use. The object-oriented paradigm provides not only a very efficient programming tool but also a very rich presentation tool in that each object can have an equivalent **presentation** object which can be visually displayed in a rather generic way. The overall conceptual architecture of the user interface is shown in Figure 19.

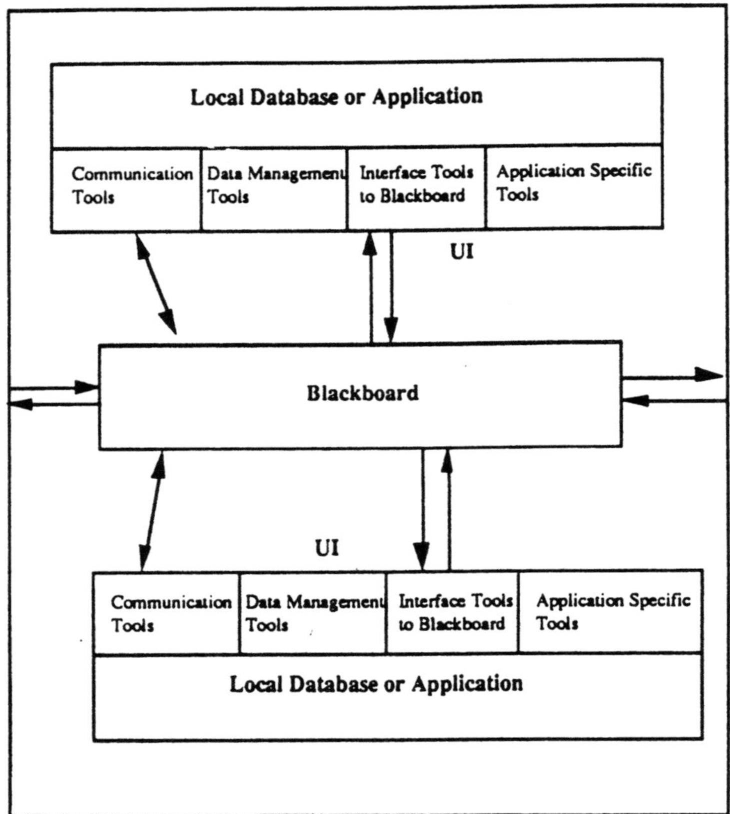

Figure 19: Overview of the User Interface

The various tools for cooperative work can be divided into four categories, as described below.

1. **Data Management Tools for the Blackboard and Local Databases**, such as:

Browsers; Editors/Displayers; Querying, Data manipulation and presentation facilities; and Documentation system.

2. **Interface Tools to Blackboard**, such as: Facilities for translation of information between Blackboard and local applications; Status checking and monitoring facilities; Communication between the Blackboard and users; and Facilities for coordination.

3. **Communication Tools**, such as: Electronic message system and Electronic conferencing facility/Negotiation tool.

4. **Application Specific Tools**, which are tailored according to the specific application. For example, an architect might have a CAD tool for designing the layout of the house. These tools are out of the scope of this project and will not be discussed further.

It must be noted that that the different components are tightly integrated, for example, editors of the data management tools are used for creating message objects of the communication system. The basic entities which the user interface acts on are all objects. How they are displayed and manipulated in the same editors/displayers depends on their structure and behavior which depends on the class they belong to.

7.2   Data Management Tools

Data management Tools are tools for creation, visualization, navigation, retrieval and manipulation of objects and their relationships in the object-base. Details of functionalities required for manipulation of these objects are described in [2].

1. **Various Graphical Browsers.** The graphical browsers can be categorized into: *composition browser, version browser, user-defined browser*, and *instance-of browser*. The composition browser depicts the compositional structure of the object base following *IS-A* and *PART-OF* relationships. The version browser shows the organizational structure in the form of *VERSION-OF* and *IS-ALT* relationships between objects. User-defined browser allows the display of objects with user defined types of relationship. Instance browser depicts the set of *INSTANCE-OF* a class. These browsers allow object hierarchies to be presented visually in two dimensions as a directed acyclic graph (DAG) in the composition browser, a tree in the version browser and a linear set in the case of the instance browser.

Powerful mechanisms are also implemented for exploring and navigating the database. By specifying the name of any object, the object will be depicted in the browser along with its related objects. At any one time, a single object (which is highlighted) is being focused on in all the browsers of a particular session; the object is known as the **Object_in_Context**. To navigate through the graph, a user has only to select another object on any browser by "clicking on it" and new graphs will be rebuilt in all browsers centered on this new object simultaneously. This mechanism which allows browsing through relationships is called *local stepping* [30], and is simple, fast and obvious. Besides, the user can also zoom in and out on part of the graph (if it is too clustered on the window) and scroll through the graphs in two dimensions. Since the number of objects to be displayed on the browsers might be too large (such that the graph is a tangled mess), a pruning mechanism can be invoked to restrict the number of objects displayed to objects that are within a specified number of levels of relationship from the **Object_in_context**. The browsers are also closely integrated with the creation, editing and querying facilities, as will be described later. The browsers, thus provide a convenient and powerful way of understanding, visualizing, communicating and finding objects in the database.

2. Object Editors/Displayers. Various types of editors/displayers provide a medium for creating, modifying and displaying objects. It is template-based in that fields corresponding to attributes are conveniently displayed for the user. Easy procedures for entering data values are provided; for example, radio buttons for choosing between alternatives and pre-defined forms for entering data are provided. Besides, any number of editors/displayers can be invoked on different objects, for example, for comparing different versions. Currently, objects are depicted in displayers and editors only in text form. Attributes which refer to another object are displayed as active icons which can invoke another editor on the referred object. Methods of objects are displayed by their name and can be invoked from the editor. Such methods include drawing operations for graphical objects or operations which manipulate their attributes. Some form of solid or geometric modeling facility will be implemented which will allow realistic graphical display of objects and their manipulation.

A multimedia editor/displayer is currently under implementation. This multimedia editor/displayer will allow the display and editing of multimedia elements. These include text, graphics, images and voice. Multimedia facilities are especially important for creating documents. In the multimedia editor, graphics, images and voice elements in an object are displayed as icons which can be invoked. When invoked, each of these elements will be displayed within a series of enclosing boxes in the editor.

3. Querying, Data Manipulation and Presentation Tools. Querying is an integral part of all database management systems. It can be seen as a more structured form of search where the nature of the desired information is known initially. Due to the nature of engineering objects, the types of queries made in engineering databases are usually more complex and varied [26], especially considering the flexibility of the object-oriented approach.

A visual querying facility which is closely integrated with the browsers is provided. It provides a very convenient way of expressing complex queries in an easy-to-understand manner. Besides, it also provides a mechanism for performing simple spreadsheet type processes (arithmetic calculations such as aggregation and multiplication) on the attributes and presentation of results in a tabular form. Examples of such queries in engineering are the calculations of amount of material used and associated costs. Facilities for plotting the query output in the form of charts can be provided too. These can be implemented as chart-plotting objects under the class of **presentation** objects.

4. Documentation system. Documents are an important part of every project, whether they are technical or non-technical. In our system, documents are another special type of objects which have the required structure and methods for document processing, for example, formatting of documents. Documents can make reference directly to other types of objects and vice versa. These documents are stored in a repository know as a **document folder**, which is simply another special type of system object. Folders can contain other folders which can store both documents and other folders. They have in-built methods for document handling such as query processing on text. Rule-based inference mechanism can be used, for example, to deduce query context in retrieving relevant documents.

Each user has a document folder and there is also a document folder in the central database which allows members of the group to access shared documents. Access rights can be specified to be restricted to certain types of personnel or individuals. The interface to the folders will be through the browsers and editors/displayers described above, since documents and folders are just another class of objects.

7.3   Interface to Blackboard

The Blackboard provides coordination and control in DICE. The Blackboard also consists of objects which are created and posted to the Blackboard by different users. The interface to the Blackboard provides the following facilities:

1.   Translation of Objects between Local Database and the Blackboard. Menu options are provided for *posting* and *retrieving* objects. The default object that is translated is the **Object_in_Context**. The *Post* function provides a way for making information accessible to other users on the system. The *Retrieve* function allows users to get information from the Blackboard. More details about these functions are provided in [22]. Users whose work might be affected by the updates will be notified whenever translations are done.

2.   Facilities for Displaying the Status of Activities of the Blackboard. One such facility provides information on the status of design. Rules can be incorporated into the system for determining design status based on the design already posted. Otherwise, a project leader can update the design status. Helpful suggestions can also be provided concerning the work of the user in relation to others. These include details on focus of tasks and pointers to required tools and data for performing the task. Graph-based display tools are provided to show the coordination and negotiation objects in the Coordination and Negotiation partition of the Blackboard.

3. Alert Messages. Alert messages due to constraints violation and dependency evaluation are used to notify users regarding conflicts in design. A *Console* window is provided where all messages are displayed automatically to notify users. On receiving these messages, the system might perform any of the following: trigger off some autonomous agents in the form of **Inference** objects, invoke the negotiation mechanism described in the next section depending on the gravity of the situation, or leave it up to the discretion of the user.

4. Coordination of Blackboard Activities. Menu options are provided for the setting of constraints and dependency information on blackboard objects. The *Project Leader* have special facilities for setting up group configuration and other coordination activities.

7.4   Communication Facilities

Communication is an important prerequisite for the success of any cooperative work. It is required for the coordination, negotiation and cooperative development of engineering ideas. Two types of facilities are provided for communication:

1. Electronic Message System. The DICE UI allows the sending of communication objects between different users. A hierarchy of communication classes is provided, each class of which has the structure and methods required for different types of messages (e.g., memo, announcement, request). These messages are created and edited using the same type of editor/displayer as for all other objects. Users can send a single message to one user, a group of users, or to a special notice board which anyone can access. Other objects can also be attached to these mail messages. There is a system object known as the **Mail_handler** which is like a personal post office with facilities for sending and receiving mail, and management of communication objects. Through the **Mail_handler**, the user is also able to specify rules for filtering and classifying messages into mail folders according to types or the semantics of the messages.

2. Electronic Conferencing System. This system allows real time conferencing between users who are geographically separated. Users would be provided with a *shared window* which displays the information of a *shared workspace* on their individual displays. The *shared window* allows users to edit and process the information of the shared workspace dynamically. A Cut and Paste mechanism provides an easy mechanism for adding new information from their private workspace or to their private workspace. Voice and video communication can also be supported. They can provide an illusion of physical presence simulating a face to face meeting. An important use of the conference system will be in **negotiation**; negotiation can be viewed as a process where goals are proposed by users or the system, for example, a new arrangement to prevent spatial conflict of two objects. A graph-based approach for displaying negotiation activities is used. The basic approach is similar to the negotiation model described in [50].

The various tools are tightly integrated through an organized and consistent structure with the use of clear menu options and multiple windows. Consistent responses, prompts and informative feedbacks are provided at every stage. The object-oriented and hypermedia approach provides a very "natural" way of representing concepts, both objects and actions. Thus, the user interface presents a user friendly environment which is conducive to cooperative work.

7.5   Related Work

Work in this area is at a preliminary stage at most research institutions. Some work on electronic conferencing is being pursued at NTT Human Interface Laboratories (Team Workstation), West Virginia University's DICE project (MONET), Olivetti Research Laboratory in Cambridge, England (Pandora), AT&T Bell Laboratories in Holmdel (Rapport), Xerox PARC (Media Space), etc.. The Object Lens project at Sloan School has some of the ingredients of our Communication module.

8   Design Agents: CONGEN (CONGEN-ST) and DATON

The various stages involved in the process of solving a typical design problem are (see Figure 20):

1. **Problem Identification.** The problem (at an abstract level), resource limitations, target technology, etc., are identified.

2. **Specification Generation.** Design requirements and performance specifications are listed; constraints and objectives are specified.

3. **Concept Generation.** The selection or synthesis of preliminary design solutions satisfying a few key constraints is performed; several alternative designs may be generated. This stage may subdivided into: 1) generate functional components, 2) obtain structures for these components, and 3) optimize structural combinations.

4. **Analysis.** The response of the system to external effects, such as loads in the case of a structure, is determined by using an appropriate model for the system. The primary purpose of this stage is to obtain the responses – preliminary and detailed - - needed to check the feasibility of a design.

5. **Evaluation.** Solutions generated during the Concept Generation stage are evaluated for consistency with respect to the specifications. If several designs are feasible then

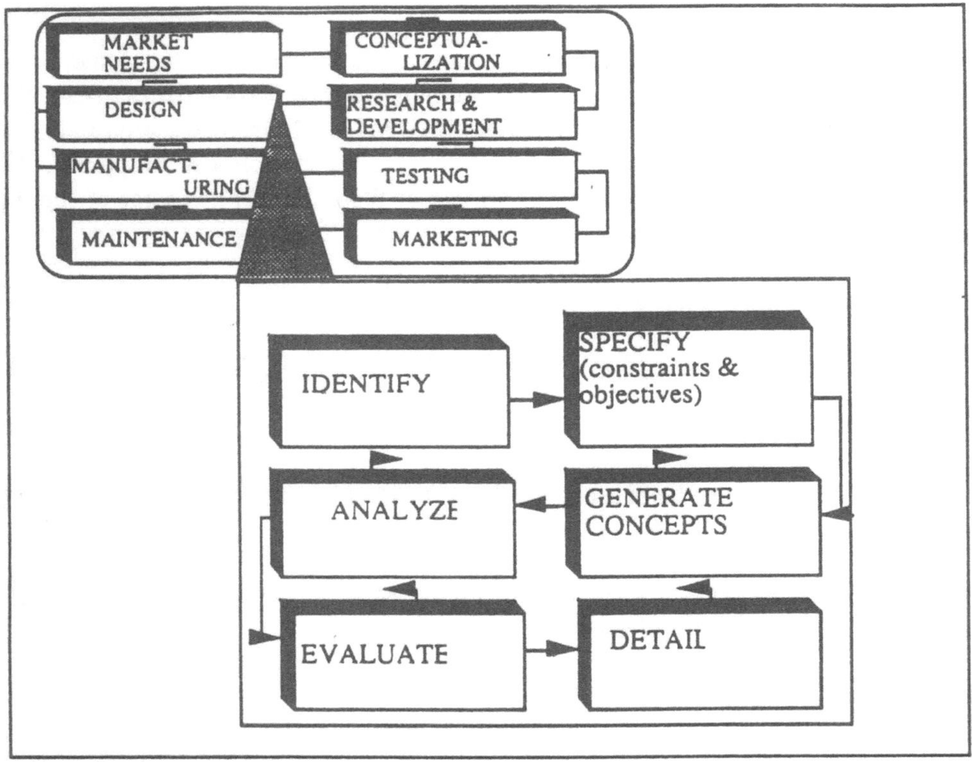

Figure 20: Design Process

(normally) an appropriate evaluation function is used to determine the best possible design to refine further. In the evaluation stage the relative optimality of several designs is determined.

6. **Detailed Design.** Various components of the system are refined so that all applicable constraints or specifications are satisfied.

There may be significant deviations between the properties of components assumed or generated at the Concept Generation stage and those determined at the Detailed Design stage, which would necessitate a re-analysis – and possibly a modification of the specifications. This process continues till a satisfactory or an optimal design is obtained. In the following sections, we will discuss two knowledge-based frameworks for design: CONGEN (CONcept GENerator), a knowledge-based shell that supports design tasks, and DATON, a detailed design system for steel design, according to LRFD specifications. In the next section, we discuss BUILDER which does construction planning. Figure 21 shows the structural engineering knowledge modules (KM's) stemming off the main DICE database.

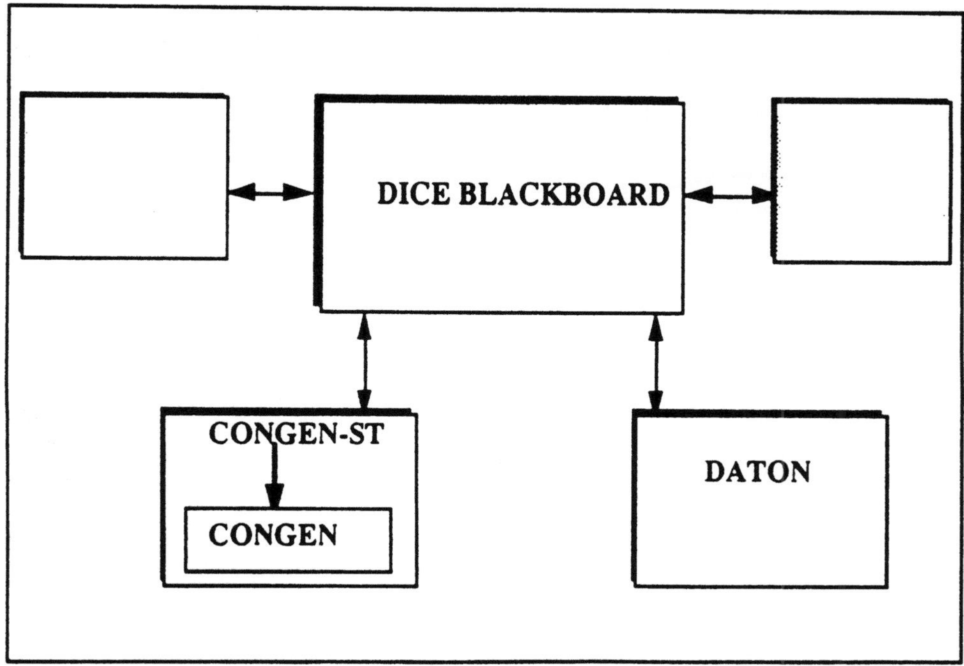

Figure 21: Structural Design Modules

8.1   CONGEN: A Knowledge-based Framework for Preliminary Design

CONGEN consists of a layered knowledge-base, a context mechanism, and a friendly user interface, as shown in Figure 22; CONGEN-ST incorporates structural engineering domain knowledge (at present buildings).

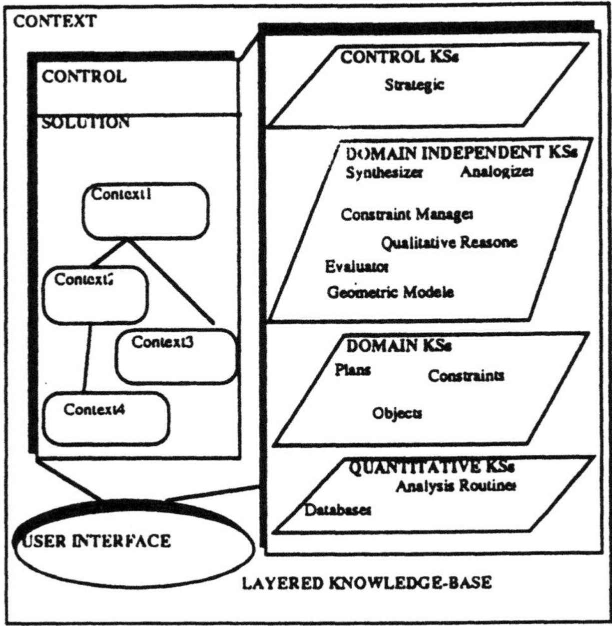

Figure 22: Schematic Overview of CONGEN

8.1.1   Knowledge-base

The Knowledge-base consists of a number of knowledge sources (KSs) that are organized into several layers or levels. The KSs that we are incorporating in CONGEN are briefly described below.

- **Strategy level KSs** determine the appropriate Domain Independent KS to fire, depending on the information provided in the Control Partition of the Context. Since this level is used to control various tasks, such as the activation of other KSs, it comprises the *task control knowledge*.

- **Domain Independent KSs [DIKSs]** perform specific tasks involved in design in a domain independent manner; DIKSs can be viewed as KBES shells. In the current implementation we are incorporating the following DIKSs.

  1. *Synthesizer* takes a set of specifications (or constraints) and generates one or more conceptual designs.

  2. *Evaluator* performs a preliminary evaluation of all the feasible alternative solutions that are generated by the Synthesizer. Evaluator acts on a network of object templates; this network exists in the Domain KS level. The root object of this network contains details of the evaluation, such as features needed for evaluation, and the evaluation function. The child nodes (or objects) represent various features; the value of each feature is determined by traversing through the alternative solution, which is represented as a tree in the Solution Partition of the Context.

  3. *Geometric Reasoner KS* is an intelligent CAD graphics system that, when implemented fully, will perform the following tasks: 1) understand engineering sketches and drawings; 2) generate geometric models and reason about these models; and 3) perform interference checking between design objects. Currently, we are implementing a non-manifold geometrical modeller (GNOMES), which can deal with multiple levels of geometric abstractions.

  4. *Constraint Manager KS* performs the evaluation and consistency maintenance of constraints arising in design. We are implementing a system – called COPLAN – which uses planning techniques to solve constraint satisfaction problems (CSPs) [19]. A planner is used as a top-level control process, guiding the search for a solution and producing an appropriate *solution plan* when the problem is solvable. The CSP is described by a *goal*. Usually the goal states which constraints should be satisfied but is more generally a list of assertions that should be true in the final world. The planner produces a non-linear *plan* at an abstract level where the different steps needed to achieve the goal are partially ordered. At the bottom level, numerical and symbolic methods are chosen in the order defined by the *plan*. The execution of a plan consists in executing the above procedure. This is very efficient in the case where one wants to vary a parameter over a certain range and to study its influence on other values for a given CSP.

- **Domain KSs** contain knowledge for a particular domain. These KSs are utilized by DIKSs. Design plans, goals, constraints, objects, and heuristic analysis procedures are some of KSs that can be incorporated at this level.

- **Quantitative KSs** contain the analytical knowledge and reference information required for analysis and design.

8.1.2 Context

The Context consists of all the solutions generated during the initial design stages (i.e., conceptual design). It is divided into two parts: the first part is the *Control Partition* that is used for storing general information; and the second part, which is the *Solution Partition*, is comprised of a tree of contexts. Multiple solutions (or partial solutions) to the design problem can be obtained from the leaf node contexts.

8.1.3 Related Work

Several computer-based frameworks were implemented for generating concepts (see [11], [41] and [51]). PRIDE/DESCRIBE (Paper path handling domain), AIR-CYL/DSPL (Air cylinders), HI-RISE/ALL-RISE (Buildings), VEXED/EVEXED (Circuits) are few examples of domain dependent/independent frameworks developed in the mid 80's; these systems used the hierarchical refinement and constraint propagation problem solving strategies.

PRIDE, implemented in LOOPS on a Xerox machine, was developed as an in-house product and it was never released to the research community; the domain independent part of PRIDE was called DESCRIBE. DSPL was a university product, developed at Ohio State University, and was widely available. This allowed researchers to experiment with DSPL in various domains. The HI-RISE project at Carnegie Mellon University evolved into EDESYN [36], which was implemented in Common LISP. Details of its usage beyond Carnegie Mellon University are not documented. ALL-RISE, which was a domain independent implementation of HI-RISE, resulted in the present CONGEN framework. VEXED, implemented in STROBE at Rutgers University for circuit design, influenced the development of EVEXED. The artifact representation used in CONGEN has several similarities to the design prototype concept developed by Gero's group [21].

In addition to the above, Stephanopoulos et al. at M.I.T. have been working an a design environment for chemical engineering for several years [49]. DESIGN-KIT and X-KIT are some of the tools developed at LISPE. DESIGN-KIT runs on Symbolics machines. X-KIT runs on IBM-PCs and is coupled with NEXPERT^{tm}. X-KIT could be used for other engineering domains, but has been geared toward chemical engineering applications.

CONGEN differs from the rest of the systems in several aspects[2]

1. **Implementation language.** CONGEN is implemented in C++/Motif, while LISP formed the base language for other systems. Hence, CONGEN can be easily integrated into engineering software.

---

[2]Please note that the current version of CONGEN does not address either innovative or creative design problem solving. Such systems are reported in [20] and [51], Volume II.

2. **User interfaces.** Considerable amount of effort in CONGEN has been spent on the user interface development. Object oriented techniques were used to implement CONGEN's user interface.

3. **Database support.** Provisions are being provided to store designs generated by CONGEN in an object oriented database (OODB).

4. **Multiple Alternatives.** CONGEN generates all feasible designs to the level specified by the user.

5. **Constraint and Geometric Modeler KSs.** The Constraint Management KS in CONGEN deals with inequalities and simultaneous equations, whereas the other systems did not deal with these adequately; a description of the constraint manager is provided in [19]. The Geometric Modeler KS provides facilities to visualize designs. The first release of CONGEN will have a variational geometry module, with a limited set of capabilities found in commercial parametric modelers such as *DesignView*[tm3].

6. **Cognitive rationale.** Many of the features incorporated in CONGEN were influenced on a case study we had conducted in the industry.

8.2 DATON

The DATON KM, like CONGEN, provides structural engineering services. However, unlike CONGEN which provides preliminary design, DATON provides full structural analysis and detailed structural design for the submitted product. The DATON sub-system is comprised of three components: a structural analysis program, a structural design program, and a controller to coordinate the flow from analysis to design. The structural analysis and design programs can be anything from academically developed software to commercially available software—as long as the input file and output file formats are understood by the controller. The controller is essentially an integrated pre- and post-processor for third party software.

Software integration is the primary issue in developing the controller. Figure 23 is a conceptual view of a session in DATON. The controller has a generic mechanism that allows the user to extract required information from the product model and to allow the information to be written into a format the analysis or design software can understand. In Figure 23, part A is the main database - the DICE system. Part B is the DATON controller. Part C is the analysis program (currently the Growltiger analysis package developed at M.I.T. or the DRAIN 2D program developed at U.C. Berkeley), and part D is the design program (currently DFRAME, an object-oriented LRFD plane frame design program, developed for

---

[3] *DesignView*[tm] is marketed by Computer Vision Corporation, MA.

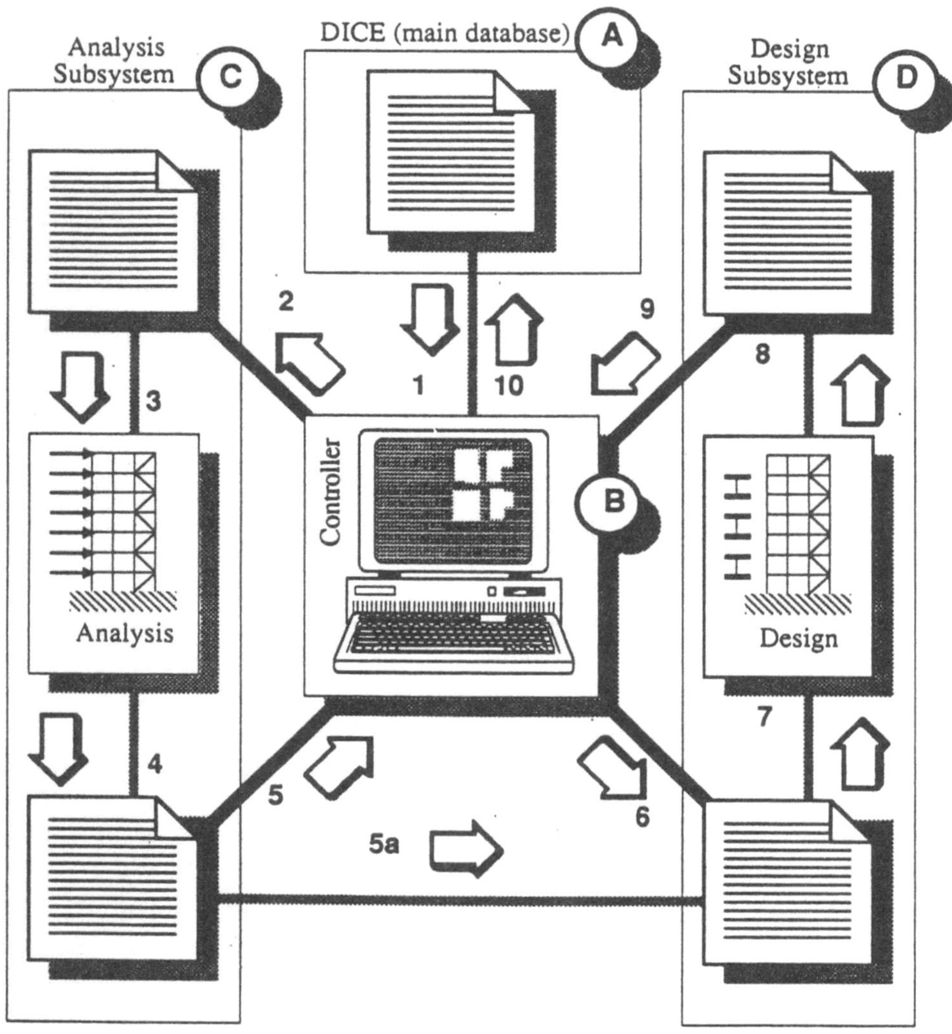

Figure 23: DATON: Flow of Information

use with DATON). The following is a description of the activities depicted in Figure 23, the flow of data in a design cycle in DATON:

1. The product data structure is passed from the DICE database to the DATON controller.

2. The controller translates the data which is in native format to the format of the analysis program.

3. The file is then submitted to the analysis program in batch mode if the analysis software has a batch mode. Otherwise, the controller executes the analysis program and assumes the user is proficient enough with the analysis software to continue.

4. When the analysis program has finished, an output file (or files) is written with the analysis results–loads, displacements, etc.

5. The controller, seeing that the analysis program has terminated, reads the necessary analysis software output file(s) and post-processes the results. The forces, displacements, resultant forces, etc. are read and translated internally into native format.

6. The controller prepares an input file in a format acceptable to the design program.

7. The design program is executed by the controller in batch mode.

8. The design program when finished produces an output file of results, which contains the designed members.

9. The controller post-processes the design programs output file(s) by internally translating the results into native format.

10. The controller resubmits the analyzed and designed product to the DICE database.

The controller should has the flexibility of allowing the user to perform the analysis procedure as many times as desired before going to the design phase.

8.2.1 Related Work

Several steel plane design systems have been developed over the years. However, most of these did not follow the LRFD strategy. One of the more recent plane frame design programs is SCAAD, which is reported in [39]. SCAAD is a menu-driven system implemented in Microsoft QuickBASIC.

DATON differs from other systems in several ways.

1. **Implementation language.** DATON is implemented in C++/Motif. The modularity provided by the object-oriented methodology, supported by C++, facilitates the incorporation of new modules with very little effort.

2. **User interfaces.** Considerable effort went into the development of the graphical user interface. Since the user interface is also implemented in an object-oriented manner, it can be easily extended.

3. **Database support.** Since DATON was developed in a much larger effort – the DICE framework – it benefits from the persistent storage mechanisms available in DICE.

4. **LRFD design specification.** DFRAME, which is the design module in DATON, uses the LRFD design specifications which is the state of the art methodology for designing steel structures.

5. **Novel design algorithms.** Several novel algorithms were developed as part of the DFRAME effort. These algorithms are reported in [1].

9   Construction Planning: BUILDER

BUILDER automates the task of generating and maintaining schedules from architectural drawings. BUILDER [9] was developed in KEE^{TM}, which is a hybrid knowledge-based programming environment. BUILDER has three major components – a drawing interface, a construction planning expert system, and a CPM algorithm – implemented as a layered knowledge-base, as shown in Figure 24. The various components of BUILDER are briefly described below.

1. **Drawing Interface.** The drawing interface layer provides for graphic input of an architectural plan. It is a menu-driven drafting system that incorporates the following features.

   (a) Provides a convenient drawing system.

   (b) Does the initial processing necessary to identify and classify the building components in a drawing, producing a representation of the drawing using a frame-based representation.

   (c) Extracts the geometric features and produces a semantic network representation of the drawing; this semantic network representation links together the frame representation of building components.

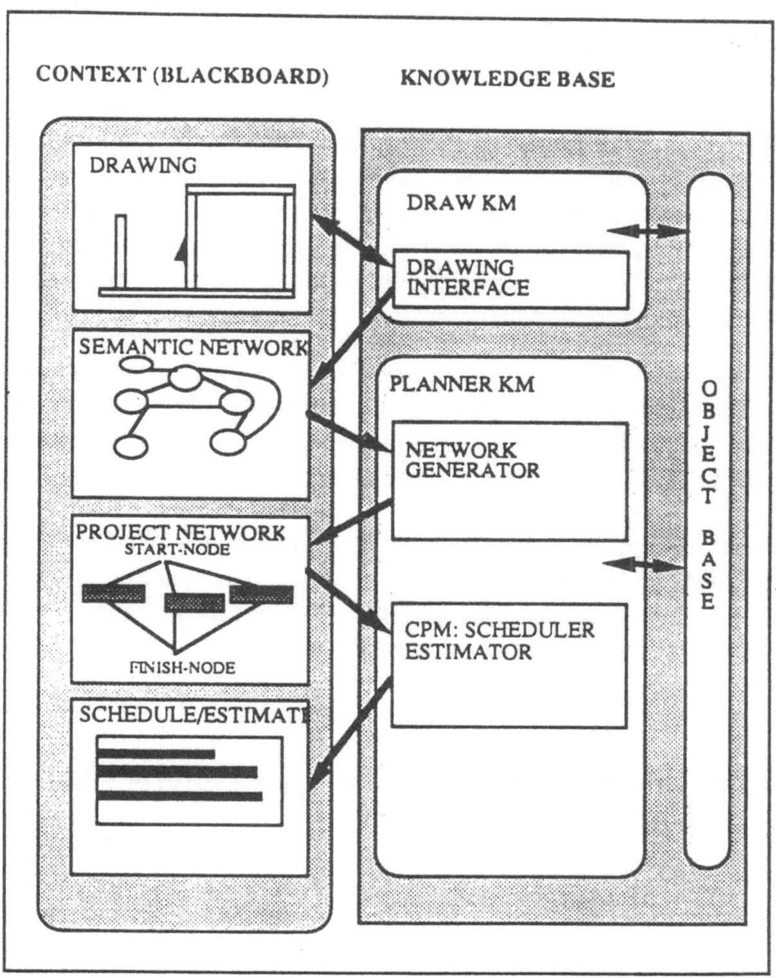

Figure 24: Schematic View of BUILDER

The friendly interface is facilitated by access to the underlying knowledge structures about building components. The menu driven system can automatically access the meanings of the symbols that it draws.

2. **Construction Planning KBES.** In an architectural drawing, the semantics of objects is normally not explicitly represented. For example there may be doors, walls, and plumbing in the drawing, but information about ordering materials for walls and doors, or having the plumbing inspected is not encoded. Neither is there any information about sequencing of tasks, or task durations, quantities, and costs. The first step in scheduling the job is to make a complete list of the tasks that need to be done. BUILDER utilizes an object-base, which is a database of engineering entities represented as frames (or objects), to complete the task list. Rules about construction methods are then activated to generate the precedence relationships between tasks. Next, BUILDER accesses a conventional database and generates an estimate of the quantities required and associated costs.

3. **CPM Algorithm.** Object-oriented and conventional CPM algorithms are implemented in BUILDER. The object-oriented approach offers some efficiency and modularity over the traditional technique in project updating, reporting, and modifying. The standard CPM algorithm is implemented for initial scheduling efficiency.

Several related systems have been developed at Stanford University's CIFE [29], [14] Carnegie Mellon University's EDRC [56], and University of Illinois, Urbana-Champaign [15]. BUILDER was one of the first planners to generate plans from engineering drawings; OARPLAN from Stanford uses similar techniques, and was developed in parallel.

10   Implementation of DICE in GEMSTONE

10.1   Overview

A prototype – called MagpieBridge – was implemented utilizing GEMSTONE, which is a commercial OODBMS. A conceptual overview of MagpieBridge is shown in Figure 25. MagpieBridge consists of two specialist KMs: an Architect KM and a Structural Engineer KM. These KMs communicate through the Blackboard. The Blackboard consists of a Critic KM in the form of a Constraint/Consistency module. The Coordination and Negotiation Blackboards do not explicitly exist in MagpieBridge; coordination is provided by the dependency objects. The various objects needed to realize the Blackboard of MagpieBridge are discussed in the following sections. Detailed descriptions of these objects and the Architect and the Structural KMs are provided in [46].

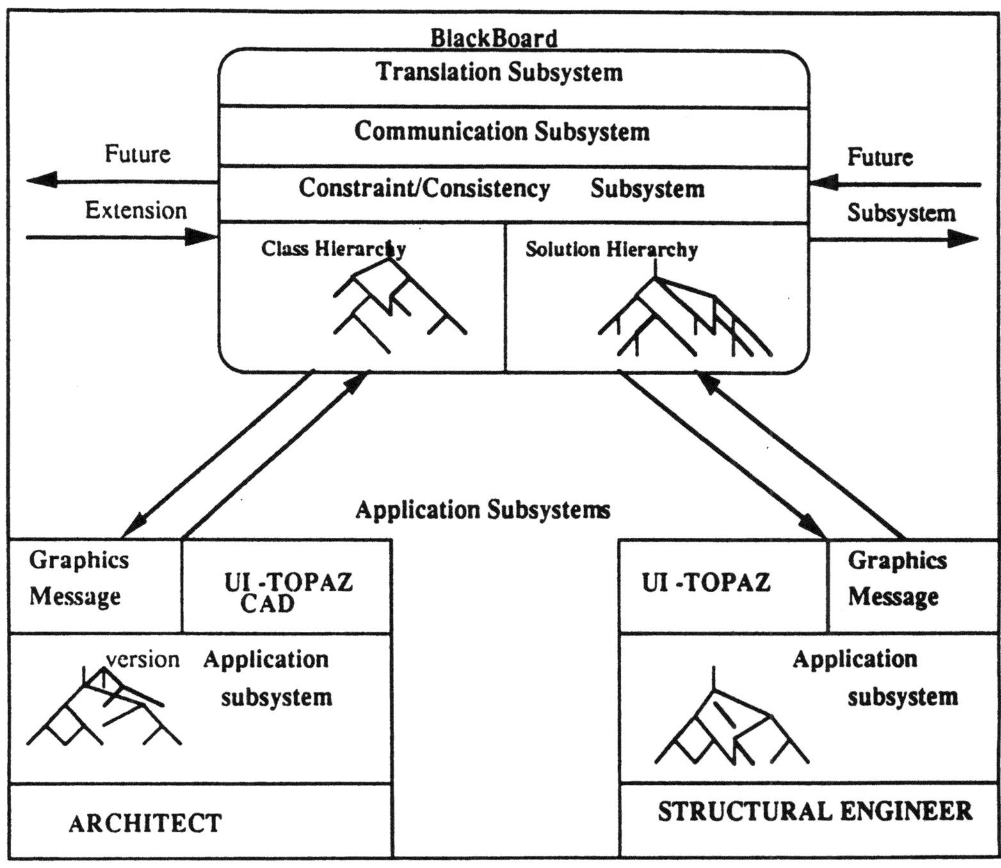

Figure 25: A Schematic View of MagpieBridge

10.2   Local and Shared Databases: GEMSTONE Dictionaries

The **UserGlobals** and the **Globals** are GEMSTONE defined dictionaries available to all users. In addition, we have identified four dictionaries: **Class_Dictionary**, **Shared_Classes**, **Project_Instan-ces**, and **Shared_Dictionary**. For all projects, the **Class_diction-ary** and **Shared_Classes** dictionaries are used by each participant. The **Class_Dictionary** stores the local classes and is private to individual users, while the **Shared_Classes** stores classes of the global database (SBB) and is shared between various users (KMs). For any particular project, two other dictionaries, **Project_Instances** and **Shared_Dictionary** maintain reference to instances of local objects and shared objects, respectively. These two dictionaries are different for each project and can be attached and detached depending on the project the user is working on.

10.3   Generic Objects

A set of objects which form the core classes of the SBB partition in the Blackboard and various KMs have been identified; A representative set of these and other domain specific objects in GEMSTONE is described in [46]. A partial taxonomy of the generic objects in the Blackboard is shown in Figure 26.

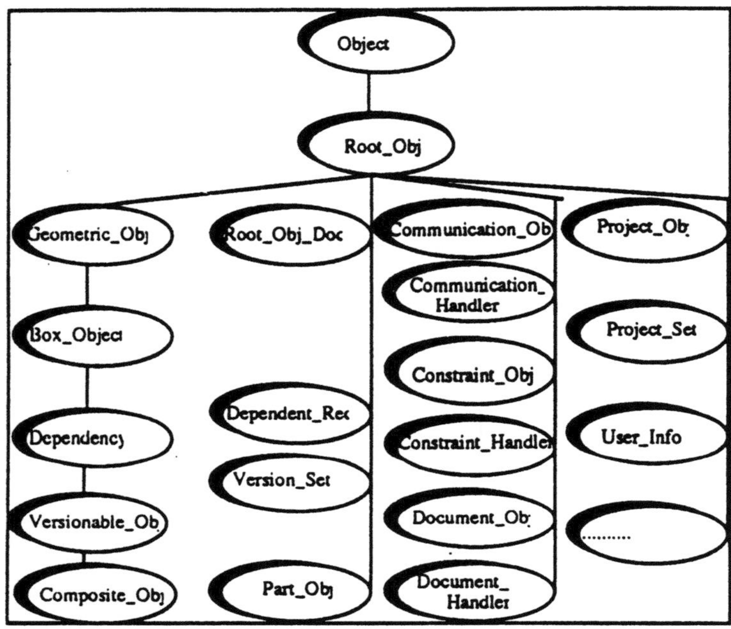

Figure 26: A Partial Taxonomy of Generic Blackboard and KM Objects

- **Root_Obj.** All classes are subclasses of this **Root_obj**. It provides timestamping, ownership stamping, and creation of reference into dictionaries.

- **Root_Objw_doc.** This is a sub_class of root object and is similar to it except that it provides a slot for attaching *document* objects, which consist of two classes: **Document_Obj**, which incorporates documentation associated with an object, and **Document_Handler**, which provides search mechanisms for document objects.

- **Geometric_Obj.** This object holds the geometric details.

- **Box_Object.** This is a subclass of **Geometric_Obj** and models an object as a box.

Box_Object has been used to describe most of physical objects in the system.

• **Dependency**. The **Dependency** object checks and keeps a record of the name and the time of various objects which have accessed a particular slot. This information is stored in the *dependents* slot in the form of a list of objects, which are instances of the **Dependency_Rec** object; note that the KM which has modified the slot will exist in the *owner* slot of the **Dependency_Rec** object. When any attribute value (slot's value) is changed a message is sent to all the KMs (or objects) which exist in the appropriate **Dependency_Rec** object. In this manner, the **Dependency_Rec** class helps in consistency maintenance.

• **Versionable_Obj**. **Versionable_Obj** objects define attributes and methods for version control. The **Versionable_Obj** object allows for the evolution of the object in the form of a version tree. A **Version_Set** object is used to hold all the versions of an instance and provides different facilities for accessing these instances based on various criteria, such as time.

• **Composite_Obj**. Most engineering objects are normally comprised of smaller components. A **Composite_Obj** class has been defined to deal with the composition of **engineering** objects. We are currently extending the capabilities of the **Composite_Obj class**.

There are a few objects - called system objects - which are used for maintaining users' information and projects' information; these objects are mostly used by domain KMs. Besides each application has a system object which provides the application specific functions and also provides interface functions to the Blackboard.

• **User_info**. This object stores the user information such as job title, name, post box for receiving communication objects and machine/display address if the user is logged on a machine.

• **Project_Obj and Project_Set**. Project objects store information about a particular project. The **Project_Set** object stores a list of all **Project_Obj** objects, i.e., a list of all projects the user is working on. The **Project_Set** object defines functions for changing the current project and loading the appropriate database.

10.4   Blackboard Subsystems Objects

The Blackboard is comprised of the following modules: a solution class hierarchy which depicts the global problem decomposition; a consistency/constraint management subsystem; a communication subsystem; and a translation/transaction subsystem. The details of these subsystems are provided in the following sections.

10.4.1   The Solution Class Hierarchy (SBB Hierarchy)

Knowledge Modules (KMs) depicting individual designers perform design operations and generate the solution in the Solution Blackboard (SBB). This solution is represented by the solution class hierarchy, which is shown in Figure 13 for our application. This global data model is different from the data models of each designer's KM, the Architect and the Structural Engineer in this example. The classes and method definitions are also different and only those attributes that are common to both application subsystems are incorporated in the classes in the central or global database (we will also refer to this as SBB). In addition, the global database classes contain attributes to store dependency information. These classes, however, do not have application specific methods, such as methods for analysis and design which exist in the Specialist KMs. The solution hierarchy is comprised of the instances of objects that are created for a particular project and posted to the central database.

10.4.2   Transaction/Translation Facilities

Transaction/Translation facilities map objects from a domain KM to that of the global database (SBB) and vice versa. These translations are not copies, as the class definitions and hierarchy are different for the different subsystems and the SBB. For the sake of simplicity and modularity, the transaction/translation capabilities are embedded in the objects and not through any external control facility. These methods are classified under the translation category. The three methods (operations) provided in the current version are: *Post*, for posting new objects onto the SBB; *Update*, for updating an existing object; and *Retrieve*, for retrieving an object from the SBB. Concurrency management is provided by GEMSTONE. In the current version, we utilized the optimistic concurrency control scheme.

10.4.3   The Constraint/Consistency Subsystem

Utilities for consistency maintenance are needed in any computer-aided cooperative work. Inconsistencies in the global database may occur if either parametric constraints or interaction constraints are violated; a parametric constraint is a constraint on a single object, e.g., *the length of the beam should not be greater than 30 units*, while an interaction constraint is a relationship between two or more objects, e.g., *the sum of the slab depth and the beam depth should not exceed 4 units*. A constraint management facility for handling interaction constraints has been implemented. This is achieved through the **Constraint_Obj** and the **Constraint_Handler** objects.

The constraint facility allows a user to specify the constraints as a symbolic relationship between objects. This is then compiled into a method which is checked when a KM accesses the SBB. The constraint system uses the *Schema Evolution* facility of GEMSTONE, which allows a method to be compiled into a class at run time. If the relationship is not satisfied, the user will be prompted either to abort the transaction or to continue. If s/he chooses to continue, then a **Communication** object will be sent to the user who has set the constraint. Thereafter, the parties can go through a negotiation phase (we are in the process of developing a negotiation framework). The constraint handler can also be invoked at anytime to check for constraint violations, if required.

10.4.4   The Communication Subsystem

Communication between members of a team is one of the most important requirement for success of their cooperative work. We have implemented a communication system which facilitates sending of communication objects or messages between different users. Two classes are defined for communication: **Communication_Obj**, wh-ich is instantiated each time a communication is warranted; and **Communication_Handler**, which manages the communication objects. Each KM will have its own communication handler. The **Communication_Handler** object has instance variables which hold the incoming and outgoing communication (**Communication_Obj**) objects. We have also implemented a concept of a post box - implemented as a **Post_box** object - which is owned by each KM and is attached to the **User_info** object. All incoming objects are received by the **Post_box** object. These communication objects are then read by the communication handler (**Communication_Handler**). The constraint subsystem uses these objects for notification.

11   Summary and Current Research Work

In this paper we have described our work on computer-aided collaborative product development. The main contributions of our work are:

1. An object-oriented blackboard architecture (DICE) that supports persistent objects.

2. An object-oriented knowledge-based building tool (COSMOS), which integrates rule-based and object-oriented programming paradigms in a C++ environment.

3. A domain independent C++-based shell for various design tasks (CONGEN).

4. An object-oriented system for detailed design of steel structures (DATON).

5. A shared information model (SHARED) for storing product information that is common to various engineering disciplines.

6. A transaction management framework that supports long duration interleaved CAD transactions.

7. User interfaces for collaborative work.

8. Prototype implementation (MagpieBridge) in a commercial object-oriented database management system (GEMSTONE).

Currently, we are working on the following components of DICE.

1. **Object-Oriented Database Support.** We will continue using commercial and non-commercial OODBMS (ObjectStore^{TM} and EXODUS [8]) as backends for DICE. We will continue our extensions on transaction management, query optimization, and version management facilities needed for collaborative work.

2. **Constraint Management.** We are adding several extensions to COPLAN. The initial version of COPLAN was developed in CLOS. We are currently porting this version to a C++ environment and integrating COPLAN with CONGEN and DICE.

3. **Design Representation in a Shared Workspace.** This work aims to provide a framework for representing product information in a shared workspace, which will provide a foundation for the development of collaborative engineering systems. We will be continuing our work on Shared Workspaces.

4. **CONGEN Extensions.** Several extensions are being made to CONGEN, including the development of a case-based reasoning framework, which will be able to store and retrieve design histories.

5. **Design Rationale.** We will be continuing our work on the CO-DRIM model.

12   Credits

The DICE project is headed by D. Sriram and Robert Logcher. Parts of this report are taken from papers co-authored with Logcher. Nicolas Groleau worked on the initial implementations of DICE. Albert Wong worked on the user interface module and is currently working on the shared workspace concept for his doctoral dissertation. Shamim Ahmed was responsible for the transaction management framework. CONGEN was a joint effort between A. Nishino, Kevin Cheong, and Parin Gandhi. Bruno Fromont and Fred Garcia are working on the constraint management system. Nestor Agbayani's SM thesis dealt with DATON. Jonathan Cherneff addressed the drawing interpretation problem in his doctoral dissertation.

He was the prime architect of the BUILDER system. Also, discussions with him helped us to reformulate and rationalize our thoughts in a more coherent manner. Navinchandra was responsible for the GHOST system and the initial work on the case-based reasoner (CYCLOPS). Feniosky Peña is currently working on CO-DRIM, the design rationale intent model. S. Gorti is responsible for the symbol to structure mapping of design alternatives. Query and storage management facilities of DICE are being implemented by Murali Vemulpati. Domain aspects (in the AEC industry) of DICE were addressed by Miriam Gross and Erik Swenson.

Funding for the DICE project comes from the IESL affiliates program and a NSF PYI Award No. DDM-8957464, with matching grants from NTT Data, Japan and Digital Equipment Corporation, USA. Partial support for Albert Wong was provided by Gleddon Postgraduate Studentship from the University of Western Australia. Kevin Cheong was supported partially by a X-Window consortium grant. Bruno Fromont's fellowship came from Aerospatiale, France.

References

[1] Agbayani, N., *DFRAME: An Object-Oriented Plane Frame LRFD Design Program with Novel Design Algorithms*, S. M. Thesis, Department of Civil Engineering, M.I.T., Cambridge, MA 02139.

[2] Ahmed, S., Wong, A., Sriram, D., and Logcher, R., A Comparison of OODBMS for Engineering Applications Technical Report No. IESL-91-03, Intelligent Engineering Systems Laboratory, MIT, Cambridge, MA 0239, 1991 [also appeared in the June 1992 issue of the Journal of Object Oriented Programming].

[3] Ahmed, S., Sriram, D., and Logcher, R., Transaction Management in OODBMS for Cooperative Product Development, ASCE Journal of Computing in Civil Engineering, January 1992 (also Technical Report Nos: IESL-90-06, IESL-91-02, Intelligent Engineering Systems Laboratory, MIT, Cambridge, MA 02139, 1990).

[4] Banerjee, J. et al., Data Model Issues for Object Oriented Systems, *Transactions of the Office Management Systems*, January 1987.

[5] Barghouti, N.; Kaiser, G. Modeling Concurrency in Rule-Based Development Environments, *IEEE Expert*, December, pp. 15–27, 1990.

[6] Barton, P. K., *Building Services Integration*, E. F. N. Spon, 733 Third Ave., NY 10017, 1983.

[7] Bjork, B. C., *Basic Structure of a Proposed Building Product Model*, Computer-Aided Design, vol. 12, no. 2, 1988.

[8] Carey, M. J., DeWitt, D. J., Richardson, J. E., and Shekita, E. J., "Storage Management for Objects in Exodus", In *Object-Oriented Concepts, Databases, and Applications*, Kim, E. and Lochovsky, F. H. (Editors), ACM Press and Addison-Wesley, 1989.

[9] Cherneff, J., Logcher, R., and Sriram, D., Integrating CAD with Construction-Schedule Generation, *ASCE Journal of Computing in Civil Engineering*, Vol. 5, No. 1, pages 64-84, 1991.

[10] Collet, C., Huhns, M., and Shen, W-M, Resource Integration using a Large Knowledge Base in Carnot, *IEEE Computer*, Vol. 24, No. 12, pages 55–62, 1991.

[11] Coyne, R., Rosenman, M., Radford, A., Balachandran, M., and Gero, J., Knowledge-based Design Systems, Addison-Wesley Book Company, 1990.

[12] Cutkosky, M., et al., PACT: An Experiment in Integrating Concurrent Engineering Systems, In *Proceedings of the First Workshop on Enabling Technologies for Concurrent Engineering*, CERC, West Virginia University, 1992.

[13] *Proceedings of the Second National Symposium on Concurrent Engineering*, Feb. 7-9, 1990, West Virginia University, Drawer 2000, Morgantown, W.Va. 26506, 1990.

[14] Darwiche, A., Levitt, R. E., and Hayes-Roth, B., ORAPLAN: Generating Project Plans in a Blackboard System by Reasoning about Objects, Actions, and Resources, *AI EDAM*, Vol. 2, No. 3, pp. 169–182, 1988.

[15] De La Garza, J. and Ibbs, W., Knowledge-Elicitation Study in Construction Scheduling Domain, *Journal of Computing in Civil Engineering*, Vol. 4, No. 2, pages 135-153, April, 1990.

[16] Eastman, C., Bond, A., and Chase, S., A Formal Approach for Product Model Information, Research in Engineering Design, Volume 2, pages 65-80, 1991.

[17] Fenves, S. J., et al., *Concurrent Computer-Integrated Building Design*, Forthcoming book, to be published by Prentice-Hall, 1992.

[18] Fenves, S. J., et al., An Integrated Software Engineering Environment for Building Design and Construction, Proceedings of the Fifth ASCE Computing in Civil Engineering, September 1988.

[19] Fromont, B. and Sriram, D., Constraint Satisfaction as a Planning Process, *Proceedings of the AI in Design Conference*, John Gero (Editor), Kluwer Academic Publishers, 1992.

[20] Gero, J. *Reports on the Creative Design Workshops*, Dept. of Architecture, University of Sydney, Australia, 1989, 1991, 1992.

[21] Gero, J., Design Prototypes: A Knowledge Representation Schema for Design, *AI Magazine*, Winter, pages 26–36, 1990.

[22] Groleau, N., *A Blackboard Architecture for Communication and Coordination*, unpublished Master's Thesis, Department of Civil Engineering, M.I.T., 1989.

[23] Gielingh, W., *General AEC Reference Model (GARM)*, ISO/STEP Technical report, IBBC-TNO, The Netherlands, 1988.

[24] Grudin, J., *Why CSCW Applications Fail: Problems in the Design and Evaluation of Organizational Interfaces*, CSCW 88, Proceedings of the Conference on Computer-Supported Cooperative Work, 1988.

[25] Gursoz, E. L., Choi, Y., and Prinz, F., *Vertex–base representation of non–manifold boundaries*, Geometric Modeling for Product Engineering, Wozny, M., J., Turner, J., U., and Preiss, K. (Editors), North–Holland, 1990.

[26] Howard, H. C. and Howard, C. S., *User Interfaces for Structural Engineering Relational Data Base*, Engineering with Computers. Vol.4, 1988.

[27] Howard, C., Abdalla, J., and Phan, D., Primitive-Composite Approach for Structural Data Modeling, *ASCE Journal of Computing in Civil Engineering*, Vo. 6., No. 1, pages 19–40, 1992.

[28] Howard, H. C., *Integrating Knowledge-Based Systems with Database Management Systems for Structural Engineering Applications*, Ph. D. thesis, Department of Civil Engineering, Carnegie Mellon University, Pittsburgh, PA 15213, 1986.

[29] Kartam, N., Levitt, R., and Wilkins,, D. E., Extending Artificial Intelligence Techniques for Hierarchical Planning, *Journal of Computing in Civil Engineering*, Vol. 5, No. 4, pages 464–477, October 1991.

[30] Katz, R., et. al., *Browsing the Chip Design Database*, 25th ACM/IEEE Design Automation Conference, 1987.

[31] Kiriyama, T., Tomiyama, T., and Yoshikawa, H., A Model Integration Framework for Cooperative Design, *Computer-Aided Cooperative Product Development*, Sriram, D., Logcher, R., and Fukukda, S., (Editors), pages 126–139, Springer Verlag, 1991.

[32] Korth, H. F., Kim, W., and Bancilhon, F., On Long-Duration CAD Transactions, In *Readings in Object-Oriented Database Systems*, Zdonik, S. B and D. Maier (Editors), Morgan Kaufmann Publishers, Inc., 1990.

[33] Korth, H. F.and Speegle, G. D., Formal Model of Correctness without Serializability, *Proceedings of ACM Sigmod International Conference on Management of Data*, 1988.

[34] Kutay, A. and Eastman, C., Transaction Management in Design Databases, In *Computer-Aided Cooperative Product Development*, Sriram, D., Logcher, R., and Fukuda, S., (Eds.) Springer Verlag, 1991.

[35] Lu, S. C-Y , *Knowledge-based Engineering Systems Research Laboratory*, Annual Report, Department of Mechanical Engineering, University of Illinois at Urbana-Champaign, April 1991, 1992.

[36] Maher, M. L., Engineering Design Synthesis: A Domain-Independent Representation, *Artificial Intelligence in Engineering, Manufacturing and Design*, 1(3), pages 207-213, 1988.

[37] Marshall, et al., R. D. , *Investigation of the Kansas City Hyatt Regency Walkways collapse*, Technical Report Science Series 143, National Bureau of Standards, Washington, D. C., May 1982.

[38] Moss, J. E. B., "Design of the Mneme Persistent Object Store," ACM Transactions on Information Systems, April 1990.

[39] Peck, B. D. and Lui, E. M., "Microcomputer Structure Member and Frame Design by LRFD," *Journal of Computing in Civil Engineering*, April 1991, pp. 141-158.

[40] Rossignac, J., O'Connor, M., *Selective Geometry Complex: A dimension-independent model for point sets with internal structures and incomplete boundaries*, Geometric Modeling for Product Engineering, Wozny, M., J., Turner, J., U., and Preiss, K. (Editors), North-Holland, 1990.

[41] Rychener, M. (Editor), *Expert Systems for Engineering Design*, Academic Press, 1988.

[42] Skarra, A. H., *A Model of Concurrency Control for Cooperating Transactions*. Ph.d Thesis, Department of Computer Science, Brown University, 1991.

[43] Sriram, D., *Knowledge-Based Approaches for Structural Design*, CM Publications, UK, 1987.

[44] Sriram, D., Logcher, R., Groleau. N., and Cherneff, J. , *DICE: An Object Oriented Programming Environment for Cooperative Engineering Design*, Technical Report IESL-89-03, IESL, Dept. of Civil Engineering, M. I. T., 1989.

[45] Sriram, D., Logcher, R., and Fukuda, S. (Editors), *Computer Aided Cooperative Product Development*, Springer Verlag, 1991.

[46] Sriram, D., Logcher, R., Wong, A., and Ahmed, S., A Case Study in Computer-Aided Cooperative Product Development, In *Computer Aided Cooperative Product*

Development, Sriram, D., Logcher, R., and Fukuda, S. (Editors), Springer Verlag, 1991.

[47] Sriram, D., et al., *Engineering Cycle: A Case Study and Implications for CAE*, In Knowledge Aided Design, Green, M (Editor), Academic Press, 1992.

[48] Sriram, D., et al., *An Object-Oriented Knowledge Based Building Tool for Engineering Applications*, IESL Technical Report, 1991.

[49] Stephanopoulos, G , et al., Design-KIT: An Object-Oriented Environment for Process Engineering, *Computers in Chemical Engineering*,Volume 11, No. 6, 1987, pp. 655-674 (See also reports from LISPE).

[50] Sycara, K. P., *Negotiation in Design*, Computer Aided Cooperative Product Development, Sriram, D., Logcher, R., and Fukuda, S. (Editors), Springer Verlag 1991.

[51] Tong, C. and Sriram, D. (Editors), *Artificial Intelligence in Engineering Design*, Three Volume Series, Academic Press, 1992.

[52] Wong, A., Sriram, D., and Logcher, R., *User Interfaces for Cooperative Product Development*, Proceedings of the Second National Symposium on Concurrent Engineering, West Viginia University, Feb., 1990.

[53] Wong, A., *Shared Workspaces for Collaborative Engineering*, Doctoral Thesis Proposal, Intelligent Engineering Systems Laboratory, Department of Civil Engineering, M.I.T., 1991.

[54] Weiler, K., *Topological Structures for Geometric Modeling*, Phd. Thesis, Rensselaer Polytechnic Institute, Aug. 1986.

[55] Zamanian, K., Fenves, S. J., and Gursoz, E., *Representing Spatial Abstractions of Constructed Facilities*, Technical Report, Engineering Design Research Center, Carnegie Mellon University, Pittsburgh, PA 15213, 1991 (To appear in a special issue of the Building and Environment journal, see also Zamanian's Ph.D. thesis).

[56] Zozaya-Gorostiza, C. and Hendrickson, C., and Rehak, D., *Knowledge-Based Process Planning for Construction and Manufacturing*, Academic Press, April 1989.

MACHINE LEARNING IN ENGINEERING DESIGN:
LEARNING GENERALIZED DESIGN PROTOTYPES FROM EXAMPLES

M.L. Maher
University of Sydney, Sydney, Australia

Abstract

The use of machine learning in engineering design should be based on a recognition of what makes design different from problem solving in general and should be guided by a representation paradigm that is useful in solving engineering design problems. Recent research in knowledge-based design has identified concept-based representation paradigms consistent with a model of the design process; this paper focusses on the representation paradigm called *design prototypes*. Conceptual clustering is a machine learning approach that provides techniques for structuring observations into generalized concepts. This paper describes how a conceptual clustering program is extended to learn engineering design knowledge by clustering function, structure, and behavior attribute-value pairs. These clusters are then used as the basis for learning associations between function, structure and behavior, resulting in generalized design prototypes.

1. Introduction

The interface between AI research and engineering design research has led to a mutual interest in the development of representation paradigms for design knowledge and design processes, as design is one of the intelligent behaviors currently associated with humans. Much of the research in AI and design has resulted in knowledge-based design systems, with some philosophical questions about the nature of design, design knowledge, and associated computational models of design. One aspect of this research has been directed towards the automatic knowledge acquisition of design knowledge bases. Machine learning techniques provide a starting point for the automatic knowledge acquisition of such knowledge bases, but currently lack a differentiation between generalized knowledge representation and generalized *design* knowledge representation.

Research in machine learning has provided theories and techniques for the development of computer programs that can learn new or modify an existing representation of domain knowledge. The various techniques available assume that the representation of the training set and the paradigm for the learned knowledge can be predetermined by the learning method rather than by the application. The results of design research in developing knowledge-based design systems provides some insight into the nature of the design knowledge that needs to be included in the knowledge base. The use of machine learning for knowledge-based design should be guided by a target representation paradigm for design knowledge and the assumption that the training set can be developed to support learning this representation paradigm. In this paper, a technique is described in which a machine learning technique is adapted and extended to fit the needs of learning engineering design knowledge. Research in knowledge-based design provides the background for selecting and extending a learning technique that can be used to learn design knowledge.

In this paper, machine learning techniques are briefly described, focussing on conceptual clustering, in order to identify a learning technique appropriate for learning generalized design knowledge. Relevant research in the representation of engineering design knowledge serves as a basis for guiding the adaptation of machine learning techniques. An extension of conceptual clustering for learning generalized design prototypes is presented. Given the design prototype as the representation paradigm to be learned, a methodology and a representation of a training set that considers the design knowledge categories function, structure and behavior is described, and a technique for learning the associations between these categories is defined. The method is illusrated by applying it to examples of trusses to learn generalised truss design prototpyes.

2. Machine Learning

Machine learning is a broad field concerned with generalizing from data and/or dynamically modifying a representation of domain knowledge. Machine learning can be considered to comprise four learning paradigms (Carbonnell 1990): the inductive paradigm, the analytic paradigm, the genetic paradigm, and the connectionist paradigm. The inductive paradigm comprises a set of techniques for automatically developing generalized concepts from examples of problems and solutions. An example of the inductive paradigm is conceptual clustering. These techniques are of interest in developing design knowledge bases for two reasons: the use of examples to describe design knowledge is easier for designers than producing a generalized body of design knowledge and recent research in design knowledge representation has focussed on the representation of design concepts rather than primarily on rules. The analytic paradigm assumes an incomplete, existing representation of domain knowledge and learns by extending the domain knowledge by solving a new problem. An example of the analytic paradigm is explanation-based learning. The genetic paradigm is based on an analogy with biological genetics and uses random search through selection and combination of exisiting solutions to find/learn new solutions. Examples of the genetic paradigm are implemented as genetic algorithms and classifier systems. The connectioninst paradigm is based on the use of a network that correlates input/output of the training set to produce a representation that can predict the output of a given new input. The connectionist paradigm is implemented as nueral networks.

In this paper, the focus is on inductive learning and how design concepts can be learned. One of the most widely studied areas of inductive learning is conceptual clustering (Michalski and Kodratoff 1990), also called concept formation. Concept formation techniques accept a training set and produce a set of clusters that group the examples in the training set. Conceptual clustering presents characteristics well suited for acquiring design knowledge. A consideration of the suitability of any machine learning technique is essential.

Some requirements for the selection, use, and adaptation of a machine learning technique for design are:

- *incremental learning:* Design knowledge is not static as there is no *correct* design knowledge. Design knowledge is updated and revised as the designer gains experience. Therefore, a machine learning technique for design must be able to accommodate new design experience by updating and modifying the generalized representation based on new examples of design solutions.
- *empirical learning:* Since there is no theory of design in a given domain, learning by experience is often guided by some general design principles rather than by specialised domain theory. For example, learning design knowledge can be guided by the need to find an association between function and structure but not necessarily guided by the specific domain knowledge such as formulas that describe the behavior of trusses.
- *learning from observation:* It is not common for an entire design solution to be classified as a positive or negative design, as is the case in some machine learning techniques. Using machine learning terminology, this implies that the learning techniques of interest employ learning from observations, that is, design situations that are not classified so that the system may learn any number of design concepts and a teacher is not needed to classify the design situations.

Conceptual clustering techniques provide a basis for learning design knowledge because they satisfy many of the requirements listed above. The major limitation of conceptual clustering techniques is that they assume that training set is described by attribute-value pairs and do not accommodate the need to consider categories of attributes.

Programs that use the conceptual clustering approach to inductive learning include EPAM (Feigenbaum and Simon 1984), UNIMEM (Lebowitz 1987), and COBWEB (Fisher 1987). Although the details differ from program to program, a conceptual clustering program accepts a set of observations as input and produces a hierarchy of clusters as output. An observation is described by a set of attribute-value pairs. A cluster is represented by a subset of the attribute-value pairs. Each cluster is also defined by the observations stored below it, and contains the observations of all its sub-clusters. When a new observation is introduced, the new observation is accommodated in the existing hierarchy using an evaluation function to determine the most appropriate cluster or to introduce a new cluster.

In COBWEB, the probability associated with each cluster and with each attribute-value pair is stored; the resulting representation is called probabilistic concept. The basis on which a concept is formed is on its ability to predict the attribute values of a new example. COBWEB is an incremental learning system in which a new example is intorduced and changes the set of concepts. When a new example is given to COBWEB, one of several operators is executed to accommodate the new observation, and the value of the category utility function is used to select between the available operators. The category utility function is based on conditional probabilities such as P(Ai = Vij|Ck) and base rate probabilities such as P(Ai=Vij); where Ai is attribute i, Vij is the j value of attribute i, and Ck is category k. The available operators include:
- classifying the object with respect to an existing cluster,
- creating a new cluster,
- combining two clusters into a single cluster (merging), and
- dividing a cluster into several clusters (splitting).

The resulting hierarchy is a probabilistic concept tree where the classification is done using a path of "best" matching nodes which depend on an object's attribute-value pairs. In contrast, in a classical hierarchy (decision tree) the decision is based on the value of a single attribute (Fisher 87). Examples given to COBWEB are restricted to lists of nominal attribute-value pairs.

BRIDGER (Reich 1990) is an adaptation of COBWEB for engineering design applications. One major difference between BRIDGER and COBWEB is that BRIDGER allows the values of attributes to be typed such as: ordered, continuous, and partially ordered. The use of multiple types of attribute's values introduces shallow knowledge about attributes. BRIDGER maintains new features in addition to the original attributes and can be viewed as an approach to constructive induction. In order to use the cluster hierarchy for design synthesis, BRIDGER accepts a new example with missing attributes and predicts the values of the missing attributes by assigning partial description characteristics based the nodes traversed.

The limitations of the conceptual clustering techniques for engineering design that are addressed here are the following.
- Observations are described by a list of attribute-value pairs and can not consider whether an attribute is a function, behavior, or structure attribute.
- Associations among attributes are not considered explicitly, but implicitly through the category utility.

In this paper these limitations are addressed by extending the application of a conceptual clustering technique, specifically by using BRIDGER, so that different categories of design attributes are considered and the associations among attributes are learned explicitly rather than implicitly through the category utilitiy. The next section describes current research in representing engineering design knowledge using knowledge-based systems representation techniques. This is followed by a presentation of a learning methodology for engineering design concepts.

3. Representing Engineering Design Knowledge

Generalizing design knowledge is particularly difficult because expert designers acquire and use their knowledge through experience. There is very little design synthesis knowledge in text books or taught in school. The result of this is that designers find it difficult to articulate their knowledge and tend to describe their knowledge through examples of design situations. This lack of a coherent body of generalized design knowledge has lead to difficulties in developing design synthesis knowledge bases that are acceptable or useful to designers.

The implementation of design knowledge bases has been largely influenced by the available representation paradigms, such as rule-based expert systems, and the nature of the design process, such as top-down refinement. Using a representation language as a guide for design knowledge representation results in a knowledge base that is stated in terms of rules, frames, objects, logic, etc. Although many design knowledge bases use these representations as a basis, the programming languages do not provide guidance in the generalizations relevant to the task of designing and the resulting knowledge bases tend to be large, complex, and difficult to update.

More recent efforts in developing knowledge-based design systems have identified appropriate process models for design and representation paradigms specific to the synthesis of design solutions. For example, Chandrasekaran describes a task oriented approach to representing design knowledge (Chandrasekaran 1990), and Maher describes various process models for design and their associated representation requirements (Maher 1990). In many cases knowledge bases for design are handcrafted by the knowledge engineer, usually resulting in generalizations that are made explicit to facilitate further knowledge base development. Two examples of this are the development of R1 followed by the development of SALT to facilitate knowledge acquisition (Marcus et al 1988) and the development of HI-RISE followed by the development of EDESYN which captures the generalized representations for synthesis used in HI-RISE without the specific knowledge about building design (Maher 1988).

Representing engineering design knowledge should be structured around an understanding of the role the knowledge plays in design rather than around a particular general purpose knowledge representation language. More specifically, the concept of design prototypes is presented as a useful representation paradigm for design knowledge. Design prototypes are introduced by Gero (Gero 1990), their implementation and application to design are described in (Gero et al 1988; Tham et al 1990).

A design prototype is a generalization of groupings of elements in a design domain which provides the basis for the commencement and continuation of a design. A design prototype represents a class of elements from which instances of elements can be derived. It comprises the knowledge needed for reasoning about the prototype's use as well as about how to produce instances in a given design context. A prototype can also be related to others either as a specialization or generalization or as a component or system to which other prototypes are the components. A hierarchy of prototypes can therefore be constructed. A individual designer's knowledge about the domain in which he/she works may be considered as being comprised of a set of prototypes, for example a structural engineer may have a prototypical understanding of how to design beams, columns, trusses, etc. Design using prototypes is a process in which suitable prototypes are sought for based on the given design specifications and are instantiated to produce instances that satisfy design goals and constraints.

What distinguishes design prototypes from general object-centered representations is the explicit classification of design knowledge into function, behavior, structure, and their relationships, to guide the synthesis of design solutions. Purely syntactic design knowledge, e.g. what a particular design looks like, is not sufficient for reasoning about generating design solutions. To facilitate reasoning about a prototype's semantics as well as syntax in design, a prototype explicitly represents function, behavior and structure, as described below.

> *Functions* are the design goals or requirements that can be achieved by using the prototype.
> *Structure* attributes describe the prototype in terms of its physical existence or the conditions for such existence. These are typically design variables whose values will be determined during the instantiation process.
> *Behaviors* are the expected reactions or responses of an instance of the prototype under the possible design environment. Performance attributes of the prototype are the behaviors of particular interest in evaluating the appropriateness and "goodness" of an instance of the prototype.

In addition to representing function, structure and behavior explicitly, design prototypes include the representation of associations between these knowledge categories. The associations that are useful for design synthesis are:
> *Function --> Structure*: These associations are primarily the heuristics accumulated

through design experience since there is no apparent function in structure or predetermined structure in function; during the early stages of design, the required functions may be known and the resulting structure is to be produced.

Function --> Behavior: A designer may use the associations from function to behavior to provide an intermediate statement of design requirements before structure is decided.

Behavior --> Structure: The associations between behavior and structure are the result of accumulated experience and analytical knowledge, providing a designer with options in generating structures that satisfy a set of behavior requirements.

The major problem with developing a knowledge base using the design prototype representation is that it is difficult to produce generalized descriptions of function, structure and behavior for a class of design objects. Design prototypes are difficult to acquire by asking a designer since designers are more comfortable describing examples of design situations. The following section introduces a methodology for generalizing a set of design prototypes from design examples using an inductive process that adapts and extends a specific conceptual clustering algorithm by considering the categories of design attributes and their role in the design process.

4. Learning Design Prototypes

Combining the techniques available in inductive learning and the use of design prototypes as the basis for representing design knowledge as function, structure and behavior, we present a methodology for learning design knowledge from observations of design situations. The methodology draws primarily on the conceptual clustering approach to machine learning. Since conceptual clustering does not accommodate the categorization of attribute-value pairs as function, structure or behavior, the methodology described goes beyond the conceptual clustering approach to include the identification of associations between categorized clusters. The clusters and their associations are used to build a generalized class of design observations, similar to the concept of design prototypes. The learning is incremental in the sense that a new observation is accommodated in the generalised representation by first updating the clusters to reflect the new observation and then updating the associations between the clusters by changing the weights of the associations relevant to the new observation.

The methodology has three distinct stages:
  1. generating clusters of design knowledge,
  2. finding useful associations between these clusters, and
  3. identifying design prototypes.
The clusters are generated by considering observations whose attribute-value pairs are categorized according to function (F), structure (S), or behavior (B). The resulting clusters

serve as an intermediate representation of the design observations. The associations between clusters that are useful for design synthesis are:

F --> S,
F --> B, and
B --> S.

The implementation of this methodology is a program called DKAO. DKAO is implemented in CommonLisp using the frame-based representation language Framekit (Nyberg 1988). A portion of DKAO uses the BRIDGER program directly. The input to DKAO is a set of design situations (design situations are referred to as observations to be consistent with the terminology introduced in Section 2) and the output is a set of design prototypes. The DKAO program will be described in three parts:

generating clusters, where the observations are clustered according to function, structure, and behavior;

determining associations, where the clusters are grouped and associations between function, structure and behavior are determined, and

identifying design prototypes, where hypothetical prototypes are identified.

4.1. Generating clusters

Conceptual clustering algorithms provide a set of techniques that accept observations as input and produce clusters as output. The particular algorithm used here is implemented as BRIDGER (Reich 1990), an adaptation of COBWEB (Fischer 1987) for engineering design applications. In BRIDGER, examples are specified as lists of attribute-value pairs, where each example has the same attributes. BRIDGER is used to consider function attributes separately to structure and behavior attributes. The clustering process remains the same, except that the result of using BRIDGER in our application is three different cluster hierarchies.

The observations, or design situations, are described by a set of attributes and values that are categorized according to function, structure, or behavior, as illustrated in Figure 1. An example of a design observation for a truss is shown in Figure 2. For each example there are three separate lists of attribute-value pairs. BRIDGER is used once for all function attributes for all examples, then for all behavior attributes, then for all structure attributes. The clusters that are the output of BRIDGER are therefore categorized according to function, structure, or behavior.

The clustering process is illustrated in Figure 3. Each observation is decomposed into three sets of attribute-values pairs; one set for each of function, structure and behavior attributes. These observations are transformed into three sets of observations, where each set represents all observations of function, structure or behavior attributes. Each set is then input to the BRIDGER algorithm to produce a hierarchy of clusters. A cluster comprises a set of attribute-value pairs that is supported by one or more observations of function attributes, behavior attributes, or structure attributes.

An example of a cluster hierarchy of behavior attribute-value pairs for truss observations is shown in Figure 4. The total number of truss observations for this hierarchy is 23. The hierarchy has three levels, the root of the hierarchy being a generalization of all the behavior attributes of the truss observations. The second level of the hierarchy has been decomposed into two sets of truss behaviors: one where the probability of a lightweight truss is higher (attribute2) and the other a heavier truss is more likely. The number in brackets next to the cluster name is the number of observations that belong to the cluster. In the actual representation, the names of the observations are stored in the cluster.

The result of generating clusters provides a starting point for identifying design concepts, but the clusters only provide classification knowledge. Determining associations between the various categories of design knowledge allows some mapping between function, behavior, and structure to be learned. These mappings capture the generalized hueristics resulting from design experience.

4.2. Determining associations for design prototypes

The design observations are considered again in order to determine the associations between function, structure, and behavior. From the hierarchy of clusters produced by BRIDGER, only those clusters that are supported by more than one observation are selected to be groups of attributes. The implication is that *generalizations* of design knowledge are based on more than one observation.

The behavior clusters shown in Figure 4 are associated with more than one observation, so for the truss observations all behavior clusters are considered as groups. Function clusters for the truss examples are shown in Figure 5, where the number of examples associated with a cluster is shown in brackets and the example names are listed below the cluster names. In the function hierarchy, F-G14 and F-G16 have only one observation so the relevant groups in this hierarchy are F-G12, F-G13 and F-G15.

The groups of function, structure or behavior attributes are used as the basis for determining associations relevant to design synthesis. The observations are used to assign weights to the useful associations. The useful associations for design synthesis are illustrated in Figure 6. An association is defined between each function, behavior and structure group and is assigned a weight based on the number of observations that support the association.

Figure 7 illustrates the calculation of a weight for an association between F-Gi and S-Gj. F-Gi has examples F-1, F-2, and F-3 associated with it and S-Gj has examples S-1, S-3, and S-5. Since obs-1 contains both F-1 and S-1, supporting the association between F-Gi and S-Gj, and obs-3 contains F-3 and S-3, also supporting the association between F-Gi and S-Gj, the weight of the association is 2; each observation contributing one unit to the weight.

The groups and their associations provide a basis for structuring design knowledge as design prototypes. The only meaningful associations are those that have a weight greater than one, indicating that the association occurred for more than one observation.

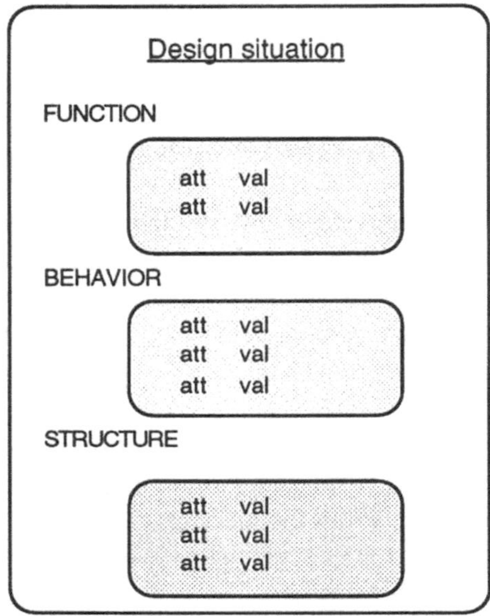

Figure 1: Representation of design situation or observation

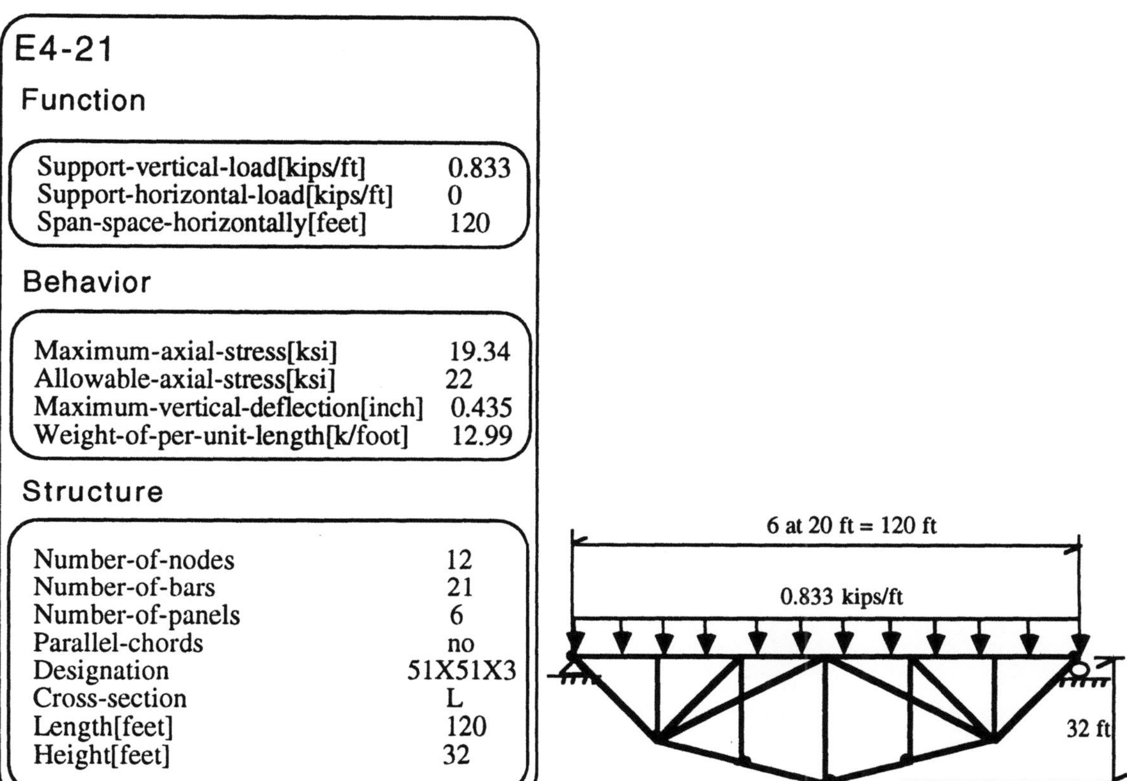

Figure 2: Example of a truss observation

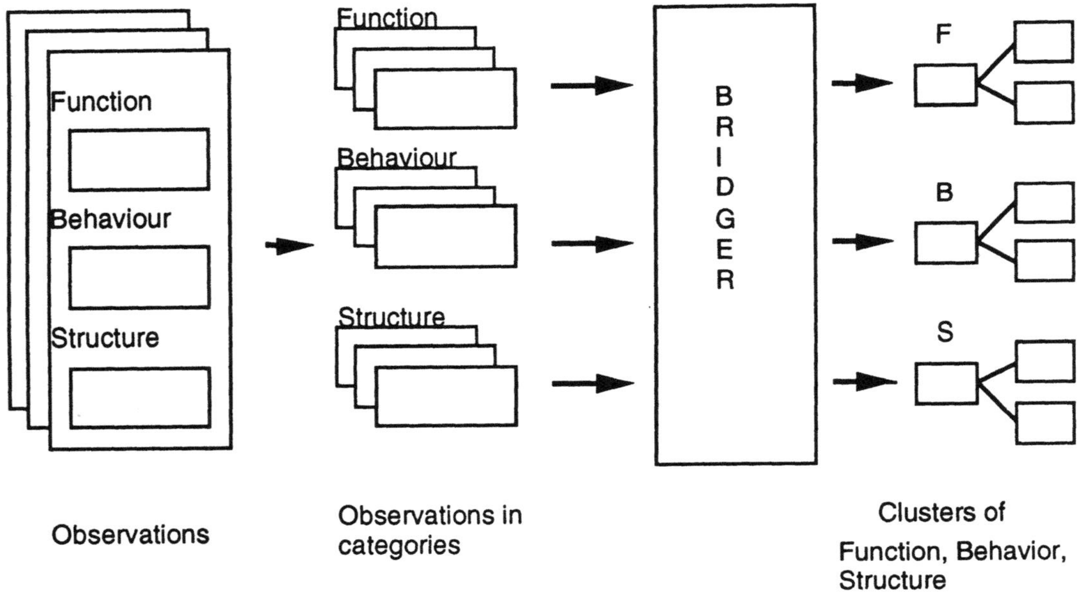

Figure 3: Clustering process

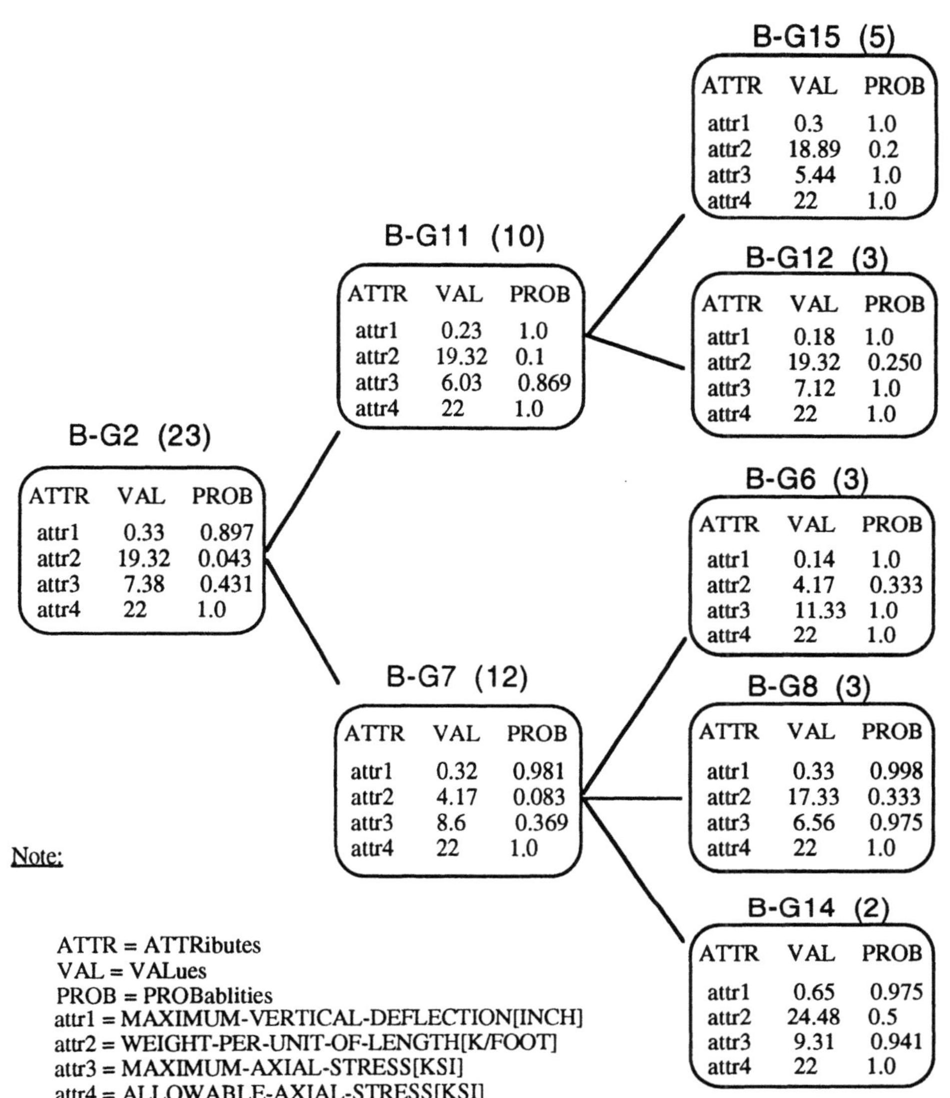

Figure 4: Cluster hierarchy for behavior attributes of truss observations (the number of examples associated with a cluster is shown in brackets)

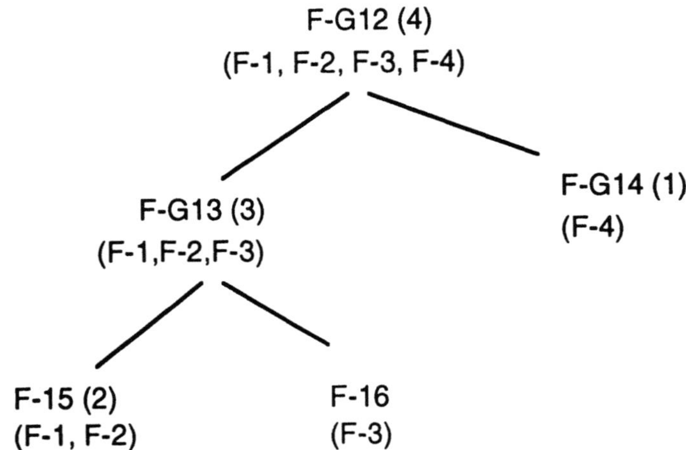

Figure 5: Cluster hierarchy of functions for bridge design (number of examples in a cluster appears in brackets, the list of examples is shown below the cluster name)

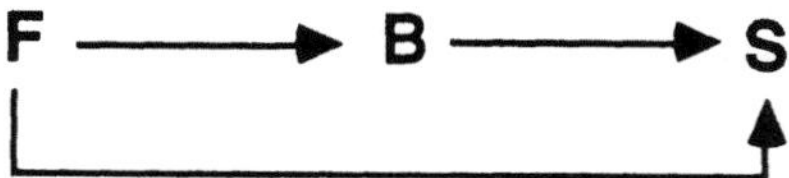

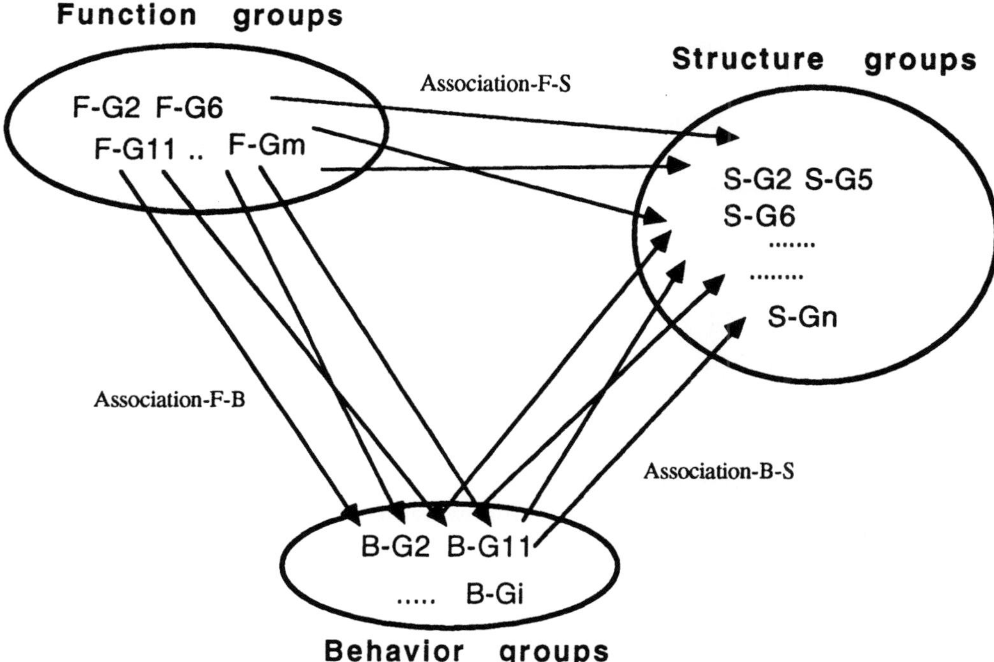

Figure 6: Associations for design synthesis

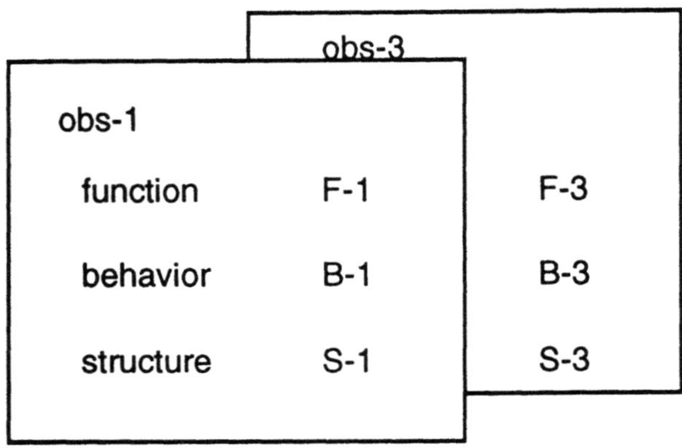

Figure 7: Example of a weight for association

4.3 Identifying design prototypes

A design prototype is determined in the following manner. The meaningful associations (those supported by more than one observation) between function and structure groups provide an initial set of partially defined hypothetical prototypes, in which only function and structure attribute-value pairs are specified. The prototypes are further defined using the meaningful associations between function and behavior groups, thereby adding a set of behavior attribute-values pairs to the definition of each prototype. Finally, the hypothetical prototypes are validated by the existence of meaningful associations between behavior and structure groups. A validated prototypes is represented by a set of function, behavior, and structure attributes where each attribute has a default value, a probability indicating how often the default has been used, and a range of values in which the prototype is valid.

Applying the methodology to learning generalized truss prototypes from the cluster hierarchies is described below.

1. Starting from a function group, F-G4, the associations to structure and behavior groups are determined from the observations. Only those associations with a weight greater than one are retained. These are illustrated below.

    Ass [F --> S]: (F-G4) = { S-G2 , S-G7, S-17, S-G16, S-G12}
    Ass [F --> B]: (F-G4) = { B-G2, B-G7, B-G9, B-G10}

2. Then the associations between each behavior group and structure groups are determined. The associations for B-G10 are illustrated below.

Ass [B --> S]:   (B-G10) = { S-G2, S-G7, S-G14, S-G18}

3. The common structure groups in associations [F->S] and [F->B] shown above are S-G2 and S-G7. These common groups validate the following prototypes.

Prototype#100   Prototype#101
Function = F-G4   Function = F-G4
Behavior = B-G10   Behavior = B-G10
Structure = S-G2   Structure = S-G7

4. The default value and probability of each attribute is determined from the clusters produced by BRIDGER. The range is determined by the examples associated with each cluster.

The result of this process is illustrated by one of the prototypes produced from the truss observations, as shown in Figure 8. The design prototype is described according to the three categories, function, structure, and behavior. Each category is described by a set of attributes, a default value and range of values, and a measure of the probability of the default value having occurred. The associations between the attributes in each category is not represented explicitly in the prototype, but implicitly by their clustering as a prototoype.

5. Results of applying DKAO to learn truss prototypes

DKAO has been applied to two sets of observations : bridge designs observations (Alem et al 1991) and truss design observations. In the domain of steel trusses, 23 observations of trusses where used. Each truss is described by the following
function attributes: support-vertical-load, support-horizontal-load, and span-space-horizontally;
behavior attributes: maximum-axial-stress, allowable-axial-stress, maximum-vertical-deflection, and weight-per-unit-length; and
structure attributes: number-of-nodes, number-of-bars, number-of-panels, parallel-chords, designation, cross-section, length, and height.

The three sets of attributes for each truss where used to produce three hierarchies, a hierarchy of functions, behaviors, and structures. The function hierarchy has 9 clusters in a 4 ply tree. The behavior hierarchy has 8 clusters in a 3 ply tree. The structure hierarchy has 10 clusters in a 4 ply tree. The interpretation of these cluster hierarchies can be made on the basis of truss classification, such as long span trusses, lightweight trusses, etc. However, this interpretation is not made explicit by DKAO.

The hierarchies were considered as groups of observations that serve as a basis for determining design prototypes through associations between function, behavior, and structure groups. The result of the 23 truss observations is 13 truss design prototypes.

Truss-Prototype13

Function

Attribute	Default value	Probablity	Range
Support-vertical-load[kips/ft]	0.93	0.93	(from 0.86 to 1.07)
Support-horizontal-load[kips/ft]	0.04	0.472	(from 0 to 0.613)
Span-space-horizontally[feet]	137.78	0.734	(from 120 to 150)

Behavior

Attribute	Default value	Probablity	Range
Maximum-axial-stress[ksi]	8.60	0.369	(from 6.24 to 10.72)
Allowable-axial-stress[ksi]	22	1.0	22
Maximum-vertical-deflection[inch]	0.32	0.981	(from 0.30 to 0.34)
Weight-of-per-unit-length[k/foot]	4.17	0.083	(from 2.31 to 48.68)

Structure

Attribute	Default value	Probablity	Range
Number-of-nodes	13	0.969	(from 12 to 16)
Number-of-bars	24	0.720	(from 21 to 29)
Number-of-panels	5	0.969	(from 4 to 6)
Parallel-chords	no	0.667	yes, no
Designation	64X64X10	0.667	51X51X3, 64X64X10
Cross-section	L	1.0	L
Length[feet]	120.0	1.0	120
Height[feet]	32.33	0.952	(from 30 to 35)

Figure 8: Truss design prototype

6. Conclusions

A method for learning design knowledge is presented that extends conceptual clustering by accommodating categories of design knowledge through the development of associations between clusters in different semantic hierarchies.

Our use and extension of inductive learning for design results in a process that produces a set of hypotheses from a set of observations. In the inductive learning method presented, design situations are observed and design prototypes are produced. The representation of design situations that serve as input to DKAO provide empirical information about the function, structure and behavior of the solutions to design problems. The resulting design prototypes are therefore heuristic generalizations of these situations.

Design prototypes can be indexed by required function, behavior and/or structure. The selection of a prototype provides a starting point for a design process. Additional knowledge is needed to analyze the performance of an instance of the prototype. The application of the learning method described in this article provides a useful starting point for the development of a knowledge base for design synthesis.

However, such a knowledge base lacks the domain knowledge needed to reason about the acceptability of a design solution. In order to incorporate such knowledge in the prototypes that result from a conceptual clustering approach, additional knowledge about the domain is needed. The prototypes are generated considering heuristic associations from function to structure, function to behavior and behavior to structure, where associations from structure to behavior represent domain theory. Such knowledge can be introduced in order to validate the resulting prototypes. This requires a shift in the machine learning paradigm from an inductive to an analytic paradigm. At this stage two techniques are of interest for the continued development of design prototype knowledge. The first one is explanation-based learning. This would result in a hybrid approach to learning design knowledge in which empirical, inductive learning is used to formulate a basic generalization of observations and domain dependent explanation-based learning is used to reason about these observations beyond the heuristic associations. The second technique of interest is genetic algorithms where the first population is composed of the prototypes that satisfy the domain theory. This approach shows promise as it enables us to introduce the design requirements as a local fitness function to evaluate the resulting prototypes.

Acknowledgements. This paper is based on a working paper developed with Leila Alem, currently at the Division of Information Technology at CSIRO in Sydney. This work was supported by a grant from The University of Sydney Special Projects Grant Scheme. The author appreciates the contribution through discussion by the staff and research students of the Design Computing Unit at the University of Sydney, Heng Li for generating the truss design examples and producing the results, and Yoram Reich for his help with the use of BRIDGER at Carnegie Mellon University.

References

Alem, L . and Maher, M.L. (1991). "Using Conceptual Clustering to Learn about Function, Structure and Behavior in Design", Kmet'91, First international Conference on Knowledge Modeling and Expertise Transfert, Sophia Antipolis, French Riviera, France, April 22-24, 1991.

Brown, D. and Chandrasekaran, B. (1985). "Expert Systems for a Class of Mechanical Design Activity", in *Knowledge Engineering in Computer-Aided Design,* (editor) J. Gero, North-Holland, pp. 259-283.

Carbonell, J.G. (1990). "Introduction: Paradigms for Machine Learning" in *Machine Learning Paradigms and Methods*, (editor) J. Carbonell, MIT/Elsevier, pp.1-10.

Chandrasekaran, (1990), "Design Problem Solving: A Task Analysis" in *AI Magazine*, Winter Issue.

Feigenbaum, E.A. and Simon, H. (1984). "EPAM-like Models of Recognition and Learning", *Cognitive Science*, 8, pp. 305-336.

Fisher, D.H (1987). "Knowledge Acquisition via Incremental Conceptual Clustering " in *Machine Learning*, 2, pp. 139-172.

Forgy, C.L. (1981). "OPS5 User's Manual", Technical Report CMU-CS-81-135, Carnegie Mellon University, Pittsburgh PA.

Gero, J.S., Maher, M.L. and Zhang, W. (1988). "Chunking Structural Design Knowledge as Prototypes" in *Artificial Intelligence in Engineering: Design*, (editor) J. Gero, Elsevier/Computational Mechanics Publications, pp 3-21.

Gero, J. (1990), "Prototypes: A Knowledge Representation Schema for Design" in *AI Magazine*, Winter.

Gluck, M. and Corter, J. (1985). "Information, Uncertainty and the Utility of Categories", in *Proceedings Seventh Annual Conference of the Cognitive Sciences Society*, Ivrine, CA. pp. 283-287.

Lebowitz M. (1987). "Experiments with Incremental Concept Formation : UNIMEM " in *Machine Learning*, 2 , pp. 103-138.

Maher, M.L. and Fenves. S.J., (1984). "HI-RISE: A Knowledge-Based Expert System for the Preliminary Structural Design of High Rise Buildings", Technical Report , R-85-146, Department of Civil Engineering, Carnegie Mellon University.

Maher, M.L. (1988). "Engineering Design Synthesis: A Domain Independent Representation" in *Artificial Intelligence for Engineering Design, Analysis and Manufacturing,* 1(3), pp. 207-213.

Maher, M.L. (1990). "Process Models of Design Synthesis" in *AI Magazine*, Winter Issue.

Marcus, S., Stout, J., and McDermott, J. (1988). "VT: An Expert Elevator Designer That Uses Knowledge-Based Backtracking". in *AI Magazine* 9(1), pp. 95-114.

McDermott, J. (1980). "R1: A Rule-Based Configurer of Computer Systems", Technical Report CMU-CS-80-119, Carnegie Mellon University, Pittsburgh PA.

Michalski, R.S. and Kodratoff, Y. (1990). "Research in Machine Learning; Recent

Progress, Classification of Methods, and Future Directions", in *Machine Learning An Artificial Intelligence Appproach Volume III*, (editors) Y. Kodratoff and R. Michalski, Morgan Kauffmann, pp.3-30.

Mitchell, T.M., Steinberg, L.I., and Shulman, J.S. (1984). "A Knowledge-based Appproach to Design" in *Proceedings of the IEEE Workshop of Principles of Knowledge-based Systems*, IEEE pp 27-34.

Nyberg, E.H. (1988). "The Framekit User's Guide Version 2.0", Technical Report CMU-CMT-MEMO, Carnegie Mellon University, Pittsburgh PA.

Reich, Y. (1990). "Design Knowledge Acquisition: Task Analysis and a Partial Implementation", in *The 5th Knowledge Acquisition for Knowledge-Besed Systems Workshop*, Banff, Canada.

Tham, K.W., Lee, H.S., and Gero, J.S. (1990). "Building Envelope Design Using Design Prototypes" in *AI in Building Design: Progress and Promise*, ASHRAE Symposium, St. Louis, Missouri.

VIRTUAL ENVIRONMENTS FOR CAD SYSTEMS

B. Feijó
Pontifical Catholic University of Rio de Janeiro, Rio de Janeiro, Brazil

ABSTRACT

Irrespective to the built-in degree of Artificial Intelligence, any tool for assisting structural and mechanics engineering design is deeply rooted into 3D environments. However, the question of 3D virtual worlds is not clearly presented in the AI-based CAD literature. This paper contributes to the understanding of Virtual Environments in CAD systems, in a way more closely aligned with the design task environment and the perceptive needs of CAD users. This paper is also a practical guide for CAD researchers and engineers who want to explore the possibilities of VE technology.

INTRODUCTION

This paper is a complement to the author's work on perception and cognition in intelligent CAD system (Feijó, 1992) presented in the Advanced School/CISM on Expert Systems in Structural and Mechanics Engineering (Tasso and Arantes e Oliveira, 1992). In the present paper, where a greater emphasis on perception is made, the author explores the use of Virtual Environments (VE) in a way more closely aligned with the design task environment of CAD users. Any attempt of building expert systems for CAD applications should consider the requirements that the new paradigm of virtual worlds imposes.

The term "Virtual Environments" is preferred to the more popular "Virtual Reality" for reasons that are clearly presented in this paper. However, the essence of the idea underlying these terms is that of interacting with virtual worlds. A general introduction to Virtual Reality (VR) can be found in the book by Vince (1995). A practical approach to the use of VR can be experienced in a number of commercial VR centers dedicated to design, engineering and manufacturing, such as the EDS Detroit Virtual Reality Center (USA). This paper summarizes the concepts of VE/VR and emphasizes their practical use. However, the focus of the work is on the perceptive aspects of VE/VR while assisting design.

The paper is organized as follows: firstly, the definition of Virtual Environments is discussed and a closer inspection into visual perception (essential for 3D design) is carried on; secondly, a series of discussions on general topics of VE technology is presented, such as the particularities of virtual senses, interfaces techniques and CAVE-like displays; thirdly, a detailed analysis of Non-immersive VE is presented, due to its importance to CAD systems; finally, the definition of design views, the concept of virtual prototyping and the practice of assisting design in Virtual Environments are discussed.

VIRTUAL ENVIRONMENTS

Visual Computing means visualizing information and exploring visual aspects of programming and human-machine interfaces. It facilitates the management of massive amounts of information, presents an intuitive graphical interface and displays models in three-dimensional form. In this regard, the author supports a more general concept, called Experiential Computing, that has been presented in recent media reports and the literature on new trends of digital technology (Cruickshank, 1996). In **Experiential Computing**, the user does more than to visualize information, that is: he/she actually experiences it. Ultimately, the user accomplishes sensory immersion in the data and the computer itself is transparent to him/her. Experiential Computing incorporates an interface paradigm called Virtual Environments.

Virtual Environments are a new interface paradigm to create a 3D world with a **virtual structure** in which the users interacts directly with virtual objects. This new interface concept is based on the senses of presence and immersion which are closely related to each other, that is: "perceiving oneself to be within a virtual structured environment is also the underpinning of a sense of *presence*, and provides a basis for identifying systems that are likely to engender a sense of immersion. Wann and Mon-Williams (1996, p.834)". As far as virtual vision is concerned, the sense of immersion is created when the user is able to perceive the computer generated image as structured in depth. Fig. 1 presents a possible taxonomy for Virtual Environments based on the sense of immersion. Formally speaking, **Virtual Reality (VR)** can be presented as a synonymous of Immersive Virtual Environments. Accordingly, Virtual Reality can also be defined as a computer-generated, three-dimensional, interactive environment in which a person is immersed.

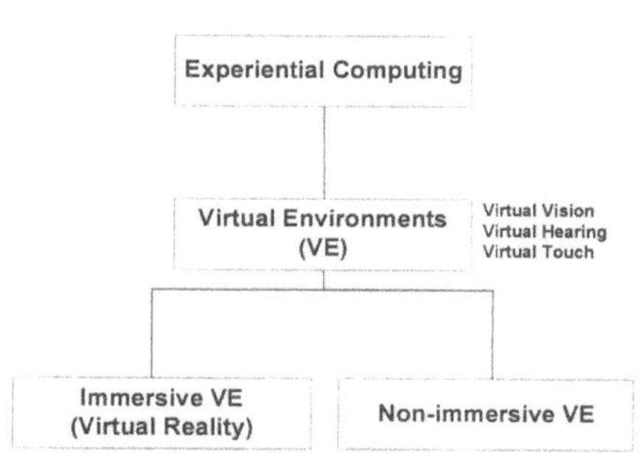

Fig. 1 Taxonomy for Virtual Environments

Virtual Reality is an effect, not an illusion. As pointed by Bryson (1996), it is the interface, not the content, that characterizes Virtual Reality. Strictly speaking, Virtual Environments are nothing more than an interface paradigm, that is: an interactive tool for the transmission, reception and manipulation of information. The search for unconstrained realism in computer interfaces makes no sense at all. For instance, irrespective to the ultimate computer of the next century, it is difficult to envisage a full digital mockup of an aircraft supporting the simulation of a real flight with all hydraulics, electrical and propulsion systems working together and being observed (or experienced) by a design team. As pointed out by Wann and Mon-Williams (1996), rather than pursuing absolute realism, simulation of 3D structures should have a clear goal in terms of information that need to be supplied and the delimits that can be placed upon the knowledge of the application domain. This is one of the reasons why the term Immersive Virtual Environments, or simply Virtual Environments, is preferred to Virtual "Reality".

3D views in AutoCAD™ and animation sequences in 3D Studio™ are not examples of Immersive Virtual Environments. No one can classify an application as an Immersive Virtual Environment (or Virtual Reality) simply because it employs 3D depiction. Wann and Mon-Williams (1996) have already pointed out that the presence of **virtual structure** distinguishes visualizations of 3D CAD models from experiences in Virtual Environments. In fact, it is easy to recognize that there is no immersion and depth perception in the examples above mentioned. However, it is difficult to set up clear criteria to distinguish immersive environments from non-immersive ones. The author claims that there is no sharp boundary between immersive and non-immersive environments, but variations motivated by a central case forming a radial category (see Lakoff (1987) for a general presentation of radial categories), as shown in Fig. 2. The central case is a full structured 3D space based on a full set of perceptual criteria P_i. and relations entangling its members. This central case corresponds to what one could name "Reality". However, the following hypothesis is made:

__Hyp__: the central case, i.e. "Reality", is close to introspection;

that is: no one is able to find a complete explanation for real perception. This hypothesis discards any attempt of pursuing unconstrained realism, which is, as a principle, undesirable and technically (and perhaps theoretically) unattainable. Therefore, one should work with subsets of P_F, that is:

$$P \subset P_F.$$

The set of perceptual criteria P works as a guideline for building good Virtual Environment systems, i.e. a VE should support at least the most salient criteria in P. Virtual Reality should support the largest number of these perceptual criteria in order to create a high degree of immersion. In contrast with VR, 3D perspective views of CAD systems are not able to make the user to perceive the computer generated image as structured in depth. In fact, the only depth cue in 3D views of traditional CAD systems is:

$$P = \{linear\ perspective\},$$

and the resulting degree of immersion is too low to characterize them as typical Immersive Virtual Environments. However, additional perception-based features in CAD environments may cause some sense of

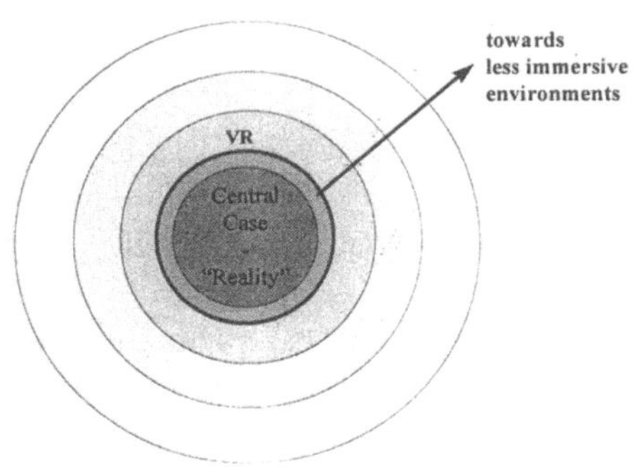

Fig. 2 Immersive VE as a Radial Category

immersion and improve 2D interfaces drastically, without the help of any annoying Virtual Reality equipment (Feijó, 1992). This latter result is particularly important for CAD systems because designers cannot wear helmets or special glasses for long periods of time in tasks where precision is at a premium.

Stereoscopic images are good examples of environments where the degree of immersion is low but the depth cues are strong enough to create a virtual environment. The contrast between situations like this one and full Virtual Reality experiences is the origin of the term "Non-immersive Virtual Environments". In Virtual Environments with low degrees of immersion it is easier to mix virtual world with real environments. In the task of assisting design, the ability of mixing virtual and real environments is very important and represents one of the reasons for the hot debate on immersive *vs* non-immersive environments for CAD systems.

One should not think that the goal of a virtual environment is always to reach full immersion (i.e. the central case in Fig. 2). Sometimes low degrees of realism in computer interfaces are mandatory, such as in the following cases: *(1)* the user is dealing with more abstract concepts; *(2)* he/she is constantly interacting with real problems in the real world; *(3)* he/she needs to reduce the distress caused by the VR devices; *(4)* the knowledge of the domain of application and the type of information to be supplied require less immersive VEs. Also low degrees of immersion are required to superimpose a virtual object to its real counterpart, as it will be discussed below in Augmented Reality. Moreover, the set of perceptual criteria P to be used by a specific system may be intentionally reduced in order to meet the goals of the intended interface. Furthermore, as a general principle, virtual environments are not constrained by the rules that govern the behavior of objects in the physical world. As mentioned elsewhere, "this is one of the attractions of the virtual setting. Wann and Mon-Williams (1996, p.845)".

DEPTH PERCEPTION

3D perception of the world is built from depth information acquired on the 2D surfaces of the eyes. The reference by McKenna and Zeltzer (1992) contains a good discussion on depth perception and 3-D display techniques. Depth cues are used to build the set of perception criteria P_d that defines the 3D virtual structure, as shown in Fig. 3, that is:

$$P_d = \{accommodation,\ convergence,\ retinal\ disparity,\ occlusion,\ shadowing,\ size,\ linear\ persp.,\ texture$$
$$gradient\ persp.,\ atmospheric\ persp.,\ optic\ flow,\ head\text{-}motion\ parallax,\ object-motion\ \text{parallax}\}$$

where "d" denotes "depth".

Ocular-motor cues come from the muscular adjustments in the eyes. **Accommodation** is the adjustment of the lens to change focus. **Convergence** is the rotation of the eyes to

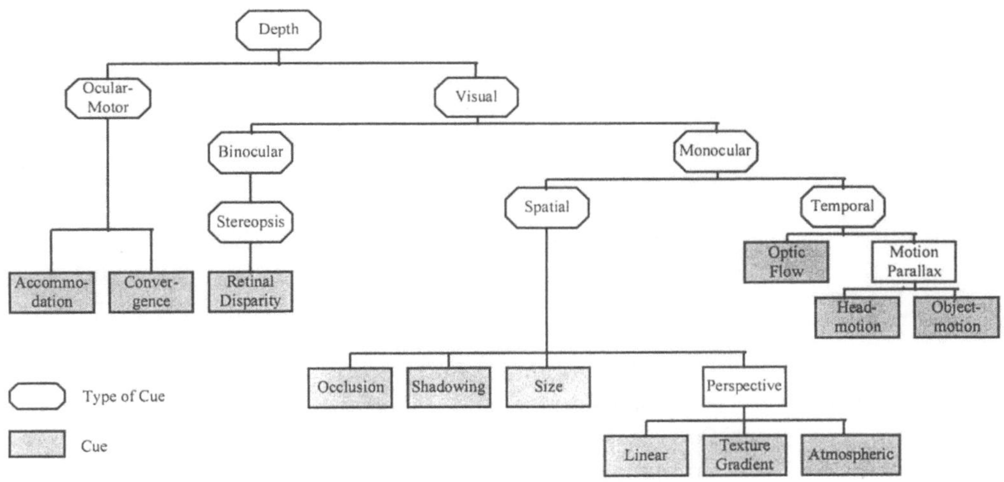

Fig. 3 Depth Cues. Adapted from Glassner (1995).

bring the point of attention (i.e. the fixation point) to fall on the central region of the retina. For long distances, ocular-motor cues contribute little to depth perception (typically the eyes muscles are completely relaxed for distances greater than 6m in the real world).

The type of visual cue provided by simultaneous processes in both eyes is based on the phenomenon of stereopsis and is called binocular depth cue. The stereopsis cue is provided by **retinal disparity** that is the relative displacement in the retinal images due to the different points of view from each eye - larger is the displacement, closer is the object. The brain interprets the retinal disparity if matches are established between both retinal images. Convergence seems to help identify useful matches for the stereopsis process.

Monocular visual cues are those extracted from a single image. As shown in Fig. 3, there are visual cues in the spatial and temporal domains. The **occlusion** cue is represented by the following rule: if object A occludes object B then A will be nearer than B. **Shadowing** refers to depth perceptions caused by the interpretation of the shadow of one object falling upon another one. The **size** cue is based on the following rules: *(1)* larger objects seem closer than smaller ones; *(2)* objects increasing their sizes seem to be moving in depth rather than physically expanding or contracting; *(3)* the distance to an object of familiar size is easily inferred.

Perspective cues refer to perceived changes of physical structures with distance. The **linear perspective** cue comes up from the convergence of parallel lines as they get farther away. The **texture gradient perspective** cue leads to depth perception by expansions/contractions of texture details and changes in relative density (e.g. spacing of people in a crowded scene). **Atmospheric perspective** (also called aerial perspective) refers to changes that arise from distant objects due to atmospheric properties, such as loss of saturation, hue shift, contour fuzziness and scattering of light through the medium.

The **head-motion parallax** cue refers to the following fact (Fig. 4): an object A nearer than the fixation point will seem to move in the opposite direction of the observer's head motion and those farther away will seem to move in the same direction of the observer's head. Because motion is relative, motion parallax also occurs when the observer is still and the objects are moving. This latter case is known as **object-motion parallax**. It is important to notice that motion parallax is a monocular type of cue and, therefore, can be tested by closing one eye. A display which provides parallax motion allows the viewer to move around the object scene. **Parallax resolution** is the number of different perspective views available to the viewer.

The **optic flow** cue concerns to optic expansions/contractions and image pattern changes over time. For instance, the optic expansion of the size of an object over time provides an estimation of arrival time.

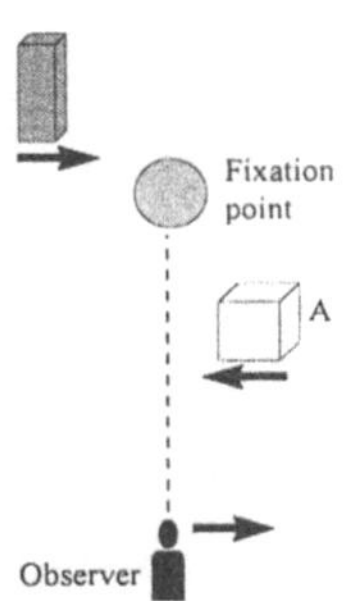

Fig. 4 Motion parallax as a depth cue

The 3D virtual structure supposes the existence of relationships amongst the elements of the set of perceptual criteria P_d, as suggested by the following possible set:

$$R_d = P_d \times P_d$$

where

$$R_d = \{< accommodation \times convergence >,\ < convergence \times retinal\ disparity >,$$
$$< optic\ flow\ x\ accommodation >,\ < motion\ parallax \times occlusion >\}$$

This set R_d should not be considered complete and further investigation is greatly required. The relations in R_d suggest that perceptual criteria should be consistent amongst themselves in a Virtual Environment. For instance, Mann and Mon-Williams (1996) have demonstrated that the presence of convergence eye movements with a display that does not promote a normal accommodative response will jeopardize the virtual structure and, consequently, the sense of immersion will be broken. In this case it is better to reduce the set P_d. As mentioned before, Virtual Environments are not constrained by the rules in the physical world. Therefore R_d is only motivated by the 3D real structure. Furthermore, the author believes, the members of R_d should be attached to "degrees of prominence" which could vary according to the interface specification.

Holography (Hariharan, 1984) seems to be the only imaging technique that can provide all depth cues, because it produces an actual 3D real image. Furthermore, although no full-parallax holographic 3D display yet exists, holography can produce images with virtually unlimited image and parallax resolutions. Real-time electro-holographic display (also named "holovideo") can produce computational holography with all of the depth cues, high image resolution and high parallax resolution found in optical holography (Lucente and Galyean, 1995). Real-time computational holography, as a marriage of computer graphics and electronic holography, is yet far from being practical (for instance, large amount of time and storage space is required to produce typical hologram 50 x 50 mm in size and rendered in hardware at a resolution of 128x64).

THE VIRTUAL SENSES

Most of the Virtual Environments are based on virtual vision. However there are some proposals for virtual hearing and touch. Smell and taste are still in the realm of science fiction.

Virtual vision requires the use of HMD (Head-Mounted Display) with typical angle ranges of 240° H (horizontal) and 120° V (vertical). In order to keep the sense of immersion, virtual vision requires rates close to 60 frames/s (while conventional film rate is 24 f/s). Some techniques have been proposed to reduce latency (i.e. response delays) based on the fact that only part of the scene needs to be updated at the maximum rate (say 60 f/s) (Regan and Pose, 1994). Strictly speaking, less than 5% of the objects needs to be updated at 60 f/s. However, the author believes, percentages greater than 5% need to be considered because of depth perception requirements.

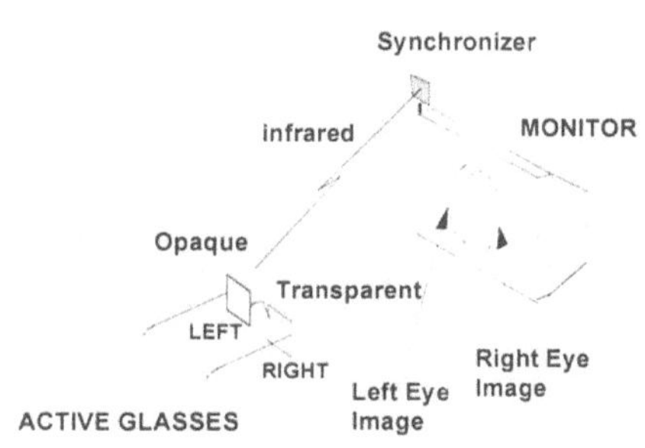

Fig. 5 Active Stereoscopic Glasses

The most popular methods to achieve stereoscopic images are illustrated in Fig. 5 and Fig. 6. The idea is to display two images on the screen corresponding to the viewpoints of each eye alternately. Therefore, in stereo mode, monitors need higher vertical rates (typically 120 Hz). Some monitors do not support this rate level at the advertised maximum resolution (e.g. a 1600 x 1280 resolution monitor may only support 120 Hz with a resolution of 800 x 640). In the active case (Fig. 5), the infrared synchronizing system makes the left and right lens transparent or opaque according to the corresponding image on the screen. In the passive case (Fig. 6), a polarizing panel is put in front of the screen in order to send the left and right images in separate planes. The passive method is quite adequate for cooperative engineering, because the images can be generated on a large screen and more than one user can experience the stereo view of the same object.

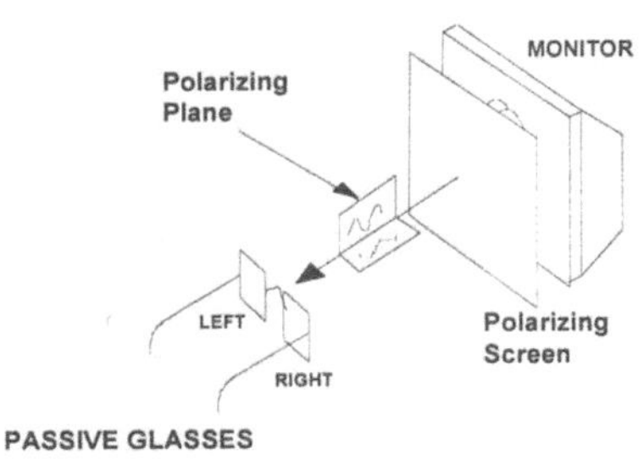

Fig. 6 Passive Stereoscopic Glasses

The high frequency of the stereo systems causes two problems: *(1)* brightness decay; *(2)* resolution decay (if bandwidth limit of the monitor is low). An important characteristic of stereoscopic images is that the latency is more tolerable than in monocular systems.

Virtual hearing is the most difficult sense to be simulated. In contrast with virtual light that is represented by particles, sound is totally represented by waves traveling in a medium with mass. The dimensions of the sound waves are similar in length to the size of the objects in scene and, consequently, diffraction, echoes and overtones are produced. Sound rendering techniques (Takala, 1992) are extremely complex and demand a lot of further research.

The stereo effect is too important for sound recognition and, consequently, it cannot be removed from virtual hearing. Stereo hearing also plays a central role in people's spatial orientation, i.e. it enhances the 3D vision structural space.

An important problem in virtual hearing is the discordance between virtual hearing and virtual vision. This type of disagreement causes the sense of immersion to be broken and may also cause physical discomfort. For this reason it is better to avoid hearing and vision coupling.

Hapitcs, the virtual senses of touch and force, has two sensorial systems: mechanoreception and proprioreception. Mechanoreception is the response of the nerves of the skin to contact, what enables people feel different types of material and vibration. Proprioreception is the muscle feedback after collision. Mechanoreception has no VE system implemented yet. Some proprioceptive systems, however, can be found in the market and academic laboratories, most of them for non-immersive VE.

SOME GRAPHICS INTERFACE ISSUES

The lessons from more than a decade of development in 2D interfaces cannot be transferred to 3D Virtual Environments straightforwardly. Genuine 3D direct manipulation and 3D widgets are still being investigated and there is a large number of open questions.

There is a clear lack of understanding of the human factors involved in 3D interactions inside Virtual Environments. Some 3D actions have intuitive support, such as grabbing and moving, but others not (such as selecting or changing properties). Additionally 3D widgets face the 3D consequences of Fitts' law, in the sense that to reach a 3D item with a pointing device takes more time than in 2D interfaces, because more space must be covered. Deering (1996) has proposed some interesting solutions for 3D widgets, such as the "fade-up" menu. Moreover, new interface paradigms have been proposed for navigation and locomotion in VE, such as the World-in-Miniature described below.

Fine motor control is much harder in 3D than in 2D because the devices (wand, gloves, ...) must be held in the space. Solutions to this problem can be found with physical supports for arms and wrists or, alternatively, through the implementation of reduction factor modes.

The type of work in a CAD environment requires numeric data enter and accurate operations. VE can incorporate hand-held bottom devices to meet these requirements, but serious fatigue problems may occur in Immersive Virtual Environments where the users cannot see their real hands and must hold the device for a long time. On the other hand, more flexible devices, such as gloves, produce inaccuracies in measurement, can cause fatigue and lack a standard gestural vocabulary (such as fist, point, ...). Therefore, glove devices should be used with low arm positions to avoid fatigue. Furthermore, appropriate visual feedback for gesture recognition should be provided. The debate on hand-held bottom devices is far from a definitive conclusion and, as mentioned by Bryson (1996), it remains to be seen if users prefer gloves or button devices.

Absolute and relative accuracy are both required during the display of virtual objects in a CAD environment. Ultrasonic trackers tend to have both types of accuracy, but magnetic ones tend to have only good relative accuracy. In most of VE systems, fixed objects tend to float in space with small erratic movements and the perceived positions suffer large distortions as the user changes head position. The perception of position stability is essential for precision work, specially when the senses of touch and force are required. Strictly speaking, the perception of position stability is fundamental for the hapitc virtual structure. Other requirements to achieve high accuracy in VE are the following: usage of high-resolution monitors with relatively flat screen; corrections of the distortions due to the curvature of the CRT; corrections of the distortions due to the index of refraction of the thick glass of the monitor; dynamic corrections for intraocular distance changes due to rotations of the viewer's eyes (Deering, 1996).

CAVE

Certainly the most impressive immersive experience is provided by the CAVE system developed by the Electronic Visualization Laboratory (EVL) of the University of Illinois, Chicago (Cruz-Neira et al., 1993). CAVE is a high-resolution environment (1280 x 512 stereo at 120 Hz, or 1024 x 768 at 96 Hz) with the capacity of mixing virtual images with real objects, in a room of approximately 3 x 3 x 2.7 m, for more than one person, with 3 projectors for the walls and one for the ceiling, as shown in Fig. 7. The CAVE has head and hand tracking systems (Ascension tethered electromagnetic sensors) to produce the correct stereo perspective and to read the position and orientation of the 3D input device (a 3-button wand). A sound system provides audio feedback (4 speakers, one in each corner of the ceiling). The users wear light weight stereo glasses (Stereographics LCD stereo shutter glasses). In the group of people immersed in the CAVE environment, only one controls the reference point of the stereo projection. The other immersed users are passive viewers.

A typical CAVE system uses two Onyx™ machines (Silicon Graphics) with 8 processors, 1 GB RAM and two graphics pipeline each. Each graphics pipeline has its own RealityEngine™ processor. In this case, each RealityEngine™ is dedicated to a specific wall. A CAVE system with the same performance might probably work with one 8-processor Onyx™ machine and two InfiniteReality™ processors.

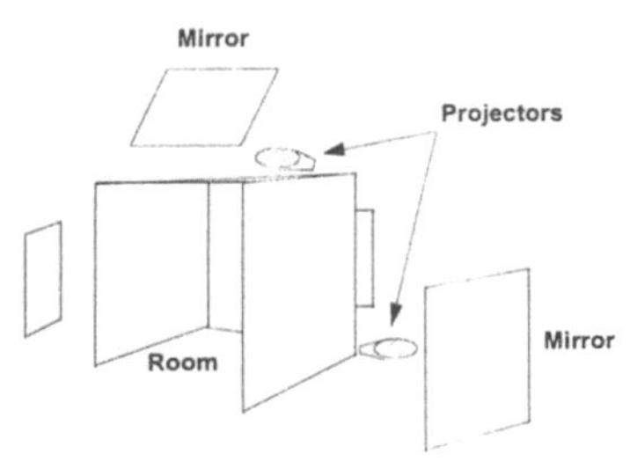

Fig. 7 CAVE

The development of CAVE has the following technical goals: large high-resolution colored stereo images with no geometric distortions; less sensibility to errors induced by head rotations; and ability to mix virtual images with real objects.

The Immersadesk™ is a spin-off of CAVE development that uses a single large display (about 1.20 x 1.5 m), rear-projected, at a drafting-table slope (45°).

NON-IMMERSIVE VE

The reference by Krueger (1991) contains a good discussion on Non-immersive Virtual Environments. This type of VE also establishes the concept of application-oriented virtual environment where the main goal is to support a specific problem-solving process. This approach, the author believes, is precisely the reason why non-immersive Virtual Environments find strong supporters in the CAD community. In this approach, the computer acts as an intelligent assistant (perhaps an expert system or even an agent system in the background) providing valuable information through multisensorial

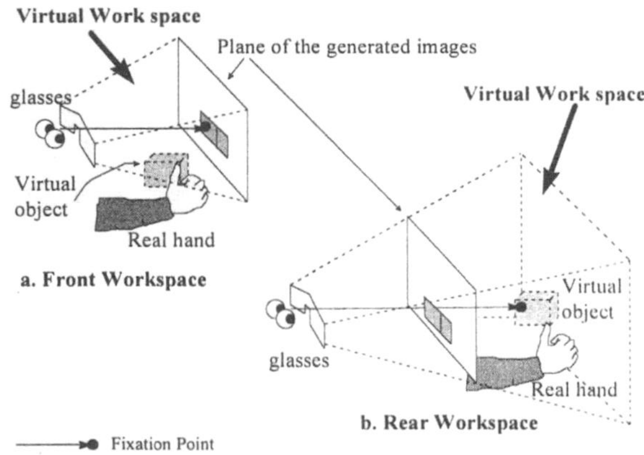

Fig. 8 VIrtual Workspace

channels of interaction. A more detailed discussion on this subject can be found in Nielsen (1993) and Marcus (1993).

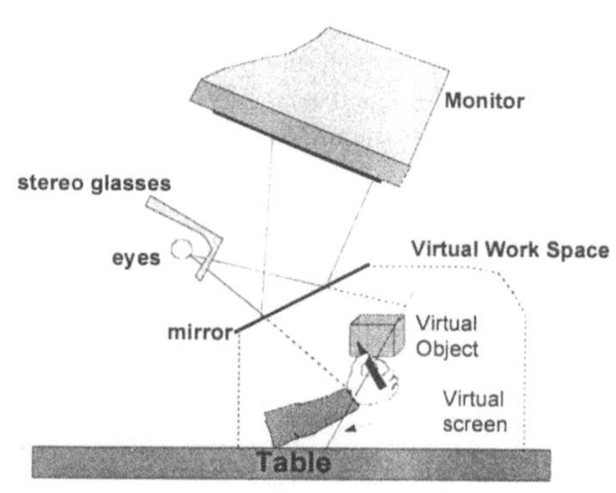

Fig. 9 Virtual Workbench (Poston and Serra, 1996)

The author claims that the central issue in Non-immersive Virtual Environments is the question of Virtual Workspace - the place where dextrous virtual work occurs. Most of the Non-immersive Virtual Environments has **Front Virtual Workspace** as illustrated in Fig. 8a. In these systems the eyes must focus on the real screen beyond the virtual object that is to be touched by the real hand. In this case, the conflict of depth cues impairs dexterity and also reduces the degree of immersion. In **Rear Virtual Workspaces** (Fig. 8b) the fixation point is the same for the virtual object and the hand. In this type of workspace, the hand-eye coordination is correct and, consequently, the degree of dexterity is quite high.

Nowadays, Rear Virtual Workspaces can only be achieved by the use of mirrors, since there is no high-resolution monitors with thin screens. Ingenious configurations have been proposed in the academic laboratories and the industry, such as Virtual Workbench and Cyberscope™. The Virtual Workbench by Poston and Serra (1996) seems to be the most comfortable and robust

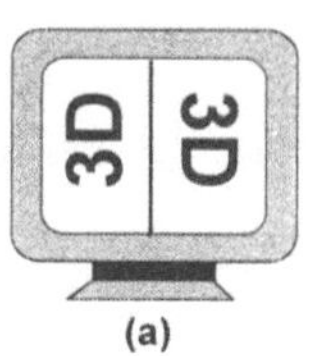

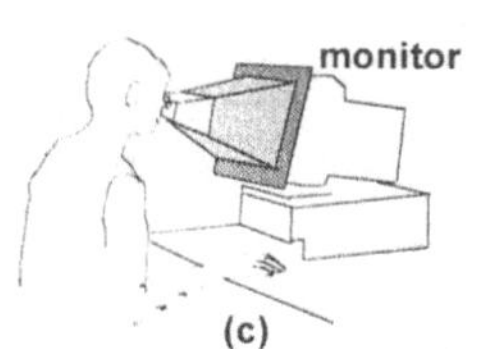

Fig. 10 Cyberscope

system (Fig. 9). The Cyberscope™ is a considerable low cost stereo device in which the user looks through a small configuration of lenses and mirrors mounted over standard monitors (Fig. 10c). In Cyberscope™ the stereoscopy pair of images are separated and rotated by 90 on the screen (Fig. 10a), then the mirror and lenses system rotates the images and produces a much wider stereo view (Fig. 10b). The size of the Virtual Workspace in Cyberscope™ is smaller than in the Virtual Workbench. Additionally, Cyberscope™ has more serious calibration problems.

The Responsive Workbench (Krueger and Froelich, 1994) is a Non-immersive Virtual Environment that uses the tabletop metaphor. The idea of the **tabletop metaphor** in VE for CAD systems is to promote the integration of the system into the working environment of the design team where it is natural the horizontal reference (Fig. 11). In the Responsive Workbench the stereoscopic images are projected over the top of a table through a system of projectors and mirrors and the stereo view is created by StereoGraphics CrystalEyes™

shutter glasses. The system tracks the head position and orientation in order to produce the correct perspective view (as in CAVE, one user is active and the remain users are passive). The interaction with the virtual world is made by gloves (for point-and-grab) and a stylus (for point-and-click) which are also tracked by the system. The problem of the eyes having their focus on a point beyond the virtual object still persists in the Responsive Workbench.

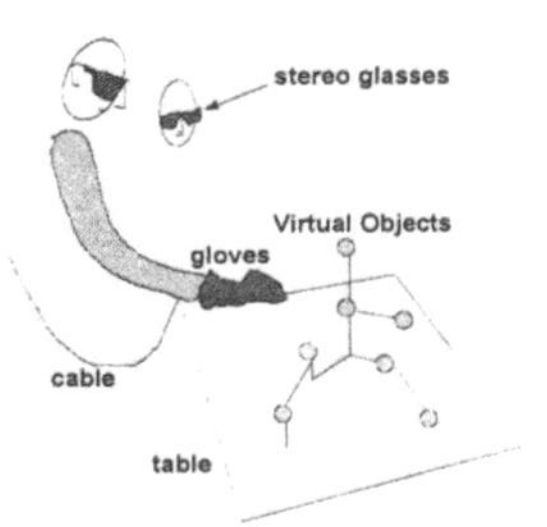

Fig. 11 The Tabletop Metaphor

Moreover, the system subjects the passive users to lurches as the active user moves his/her head. However, the commercial version of the Responsive Workbench (Immersive Workbench™, Fakespace) seems to be a highly promising solution for cooperative engineering around a drafting table. Moreover, the Responsive Workbench is a good starting point to explore new trends in HCI (Human-Computer Interface) where traditional workstations are substitute by tracking systems, cameras, projectors, microphones, speakers, and where voice and gesture recognition substitute conventional mouse and keyboard interactions.

Haptic interfaces are more adequate for Non-immersive VEs, because of the small Virtual Workspaces usually associated to this type of VE. Hapitc systems for precision work in CAD environments have even a smaller workspace (about 20 x 27 x 38 cm). In order to meet the requirements of CAD work, hapitc systems should offer an I/O library and a programming language to develop specific procedures. The system PHANToM™ (with C++ IOLIB and GHOST™ programming toolkit) has been used in academic research and industrial applications.

A remarkable innovative approach to Non-immersive Environments is the concept of **Augmented Reality**. Augmented Reality uses see-through HMDs that let the user see the real world around him/her. The central idea in Augmented Reality is to enhance the user's perception and interaction with the real world. A recent use of Augmented Reality to guide a technician in building wiring harness systems in the aircraft industry has been reported by Sims (1994). Maintenance applications have been investigated elsewhere (Feiner et al., 1993). The basic problem in Augmented Reality is the registration problem, that is the difficulty of making the real and virtual objects to be aligned to each other. Registration errors cause the sense of coexistence to be broken. Latency, one of the sources of registration errors, makes the virtual objects appear to "swim around" the real objects.

ASSISTING DESIGN IN VE

In this paper, the author claims that the use of virtual worlds in intelligent CAD systems is required by the following cognitive characteristics of the design processes: the set of invariants that delimit the design activities; and the characteristics of the design task environment. These invariants and characteristics were firstly presented by Goel and Pirolli (1989) and subsequently adapted for solid modeling processes by Feijó (1992). In this context, one can state that Virtual Environments in CAD systems are supported by the following invariants: *(1) Limited or delayed feedback from the real world* (therefore, the designer must experience virtual feedback), *(2) Autonomy of artifact* (hence the designer must predict all necessary interactions), *(3) Temporal separation between specification and delivery* (in this case, the temporal separation enables the designer to have virtual experiences); and *(4) Costs associated with every action* (consequently the designer is forced to anticipate as many consequences of an action as possible).

Once the cognitive needs for Virtual Environments are proved, there are two additional points to be considered: *(i)* the perceptive needs of the designer should be satisfied by the CAD systems; and *(ii)* CAD system architectures should integrate 3D virtual objects with other types of objects which exist in an integrated design environment. The first point is discussed in the previous sections of the present paper. The second point requires the concepts of design views and virtual prototyping. A **design View** is any set of design objects and their links. In this paper, the collection of all design views is called **Design Space**. Virtual Environments are integrated into the Design Space through a special design view called Geometric View which is, essentially, the Modified CSG Tree presented by Feijó (1992). CAD expert systems or any other AI-based CAD application are represented by design views that are orthogonal to the geometric view, as illustrated in Fig. 12.

The key concept in **Virtual Prototyping** is to build a full virtual artifact (called digital mock-up) in such a way that design and manufacturing problems are anticipated and discussed in a cooperative

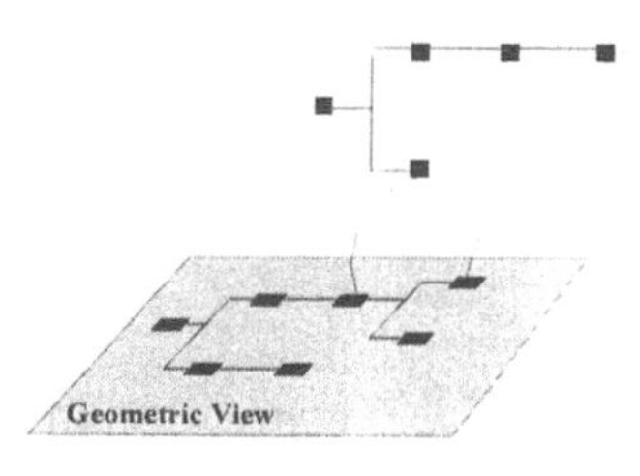

Fig. 12 Design Views

environment. A full digital mock-up is more than a common reference to the design team; it is the place where several design assistance tools interact with the design team. Strictly speaking, a digital mock-up is a design view.

The possibilities of VE in assisting cooperative design processes are better presented with the help of the integrated CAD system architecture shown in Fig. 13. In this architecture, the Integration Bus is based on an intranet model and a standard Geometry Bus. The **Integrated Digital Mock-up** is the heart of the integration architecture because all product data are associated to the virtual objects in the 3D model. In this case, any virtual object has geometric attributes, design intent attributes, manufacturing attributes, cost attributes, part number references, document references and other attributes. The geometric data in the intranet is supported by a Geometry Bus which consists of a standard object-oriented geometric library and a standard geometric data format, such as those found in ACIS™. Web applications implement a version control system and support a distributed object architecture in accordance to CORBA standards. Any user logged in the intranet can build his/her Design Space requesting a copy of a part of the digital mock-up. Several design assistant tools can be evoked as Java applets and used to manipulate the design space.

The Integrated Digital Mock-up is built in parallel with the database tables of the Product Structure. The STEP file is created from the data in the Integrated Digital Mock-up and the Product Structure. The Product Structure contains all information required to produce the product and to maintain it, including part lists, revision orders, references to 2D drawings, references to 3D objects in the mock-up, fabrication orders, references to manufacturing CN programs and other data items.

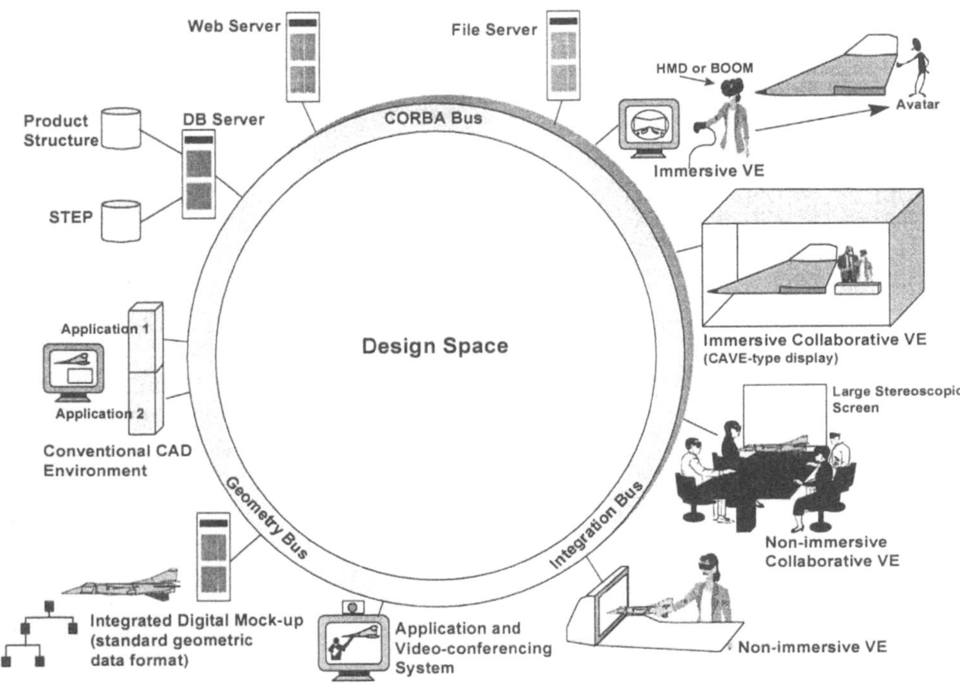

Fig. 13 Integrated CAD System

Three-dimensional visualization that are not very intense can be made with the help of VRML (Virtual Reality Modeling Language) files requested from the intranet. The VRML files can be created from the mock-up files or the STEP files. This sort of visualization is particular useful for application and video-conferencing systems.

The issue of immersive *vs* non-immersive VE in CAD systems is not settled down in the CAD community yet. Most of the non-immersive VE displays use standard workstations in stereo mode and render much higher-resolution views of an artifact than those produced by immersive VE systems. Furthermore, in non-immersive VE systems the user sits in a comfortable chair and avoids the fatigue of standing up with no support for arms and hands. The supporters of non-immersive VE claim that immersive HMDs prevent easy access to standard design tools such as paper documents and other designers in the room. Also they point out navigation problems in immersive VE. For instance, the designer takes a considerable amount of time to move around large virtual prototypes which are in the same scale of the user. Another example is that the path to a new location in an immersive VE may require planning to avoid obstacle and a lot of effort is spent to travel through obstacles. On the other hand, the supporters of immersive VEs claim that the above-mentioned problems will soon be overcome by higher-resolution HMDs and better 3D interfaces. For instance, the World-in-Miniature paradigm (Stoakley et al., 1995) seems to be a natural "idiom" for Immersive Virtual Environments that solves most of the navigation and locomotion problems.

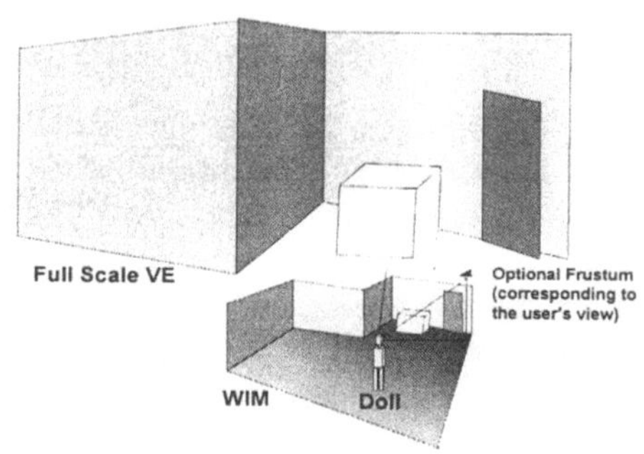

Fig. 14 World-in-Miniature interaction technique

The use of World-in-Miniature (WIM) in Immersive Virtual Environments has a great potential for applications in CAD systems. The WIM is a hand-held miniature 3D map with an anthropomorphic doll icon that represents the user, as shown in Fig. 14. When the user manipulates the doll in the map, the corresponding view in the full scale VE is simultaneously update. Locomotion problems are reasonably solved with multiple WIMs in the same scene. As pointed by Stoakley et al. (1995), each WIM acts as a portal onto a different, perhaps distant, part of the surrounding immersive world, or to a different world in a completely different context. In order to avoid the user to shift focus from the miniature back to full scale, the system animates the user into the miniature. The idea is that the user becomes the doll. Stoakley et al. (1995) speculate that users mentally envision themselves to be at the doll's vantage point, a much stronger association that merely using the miniature as a symbolic representation for viewpoint. The World-in-Miniature paradigm is more efficient and robust than techniques that allow the user to scale the virtual world down to a miniature, select a new vantage point, and then re-scale the miniature back up, as in the virtual wind tunnel by Bryson and Levit (1992).

The author claims that both immersive and non-immersive VE should be used in an integrated CAD system. Good immersive examples are: avatar incorporation systems, BOOM-type displays for engineering simulation and CAVE-type systems. Immersive VE where the designer incorporates an avatar (i.e. a virtual copy of the user in the VE) is quite useful to test human factors. For instance, a designer can easily check visibility and control accessibility in the virtual cockpit of an airplane. The use of virtual humans to assist design does not suppose that the user's movements are mapped one-to-one. The idea is to have intelligent multipliers so that the suggestion of an action is enough to precipitate complete behaviors in the virtual human (Badler et al., 1993). Virtual humans are also useful in conventional CAD environments in a number of ways, such as: crash simulation; workplace assessment; human strength analysis; design of instrument panels; check of maintenance procedures (specially in areas of difficult access). These applications in conventional CAD environments should be mixed with experiences in immersive VEs in order to produce a better design in a shorter period of time.

Heavy engineering simulations in immersive VEs are accomplished with BOOM-type and CAVE-type displays. BOOM (Binocular Omni-Orientation Monitor) technology (Bolas, 1994) is several times cheaper than CAVE-type technology and it has been used by a number of companies. The Virtual Wind Tunnel presented by Bryson and Levit (1992) is a well-balanced example of the use of CRT-based BOOM display and glove interface. In this application the visualizations include streamlines, isosurfaces, cutting planes, numerical values and color variation according to air density.

Immersive VE for collaborative engineering is only achieved today with CAVE-type displays. The automotive industry has approved the use of CAVE-type technology for styling and design review (Ellis, 1996). For instance, in styling stages, designers' sketches are quickly converted into stereo scenes to provide a quick sense of the spaciousness of interiors or the quality of the external surfaces. In design review stages, immersive VEs can solve communication problems for design teams, which include people of diverse backgrounds and skills, such as designers, engineers and manufacturing engineers.

Non-immersive VE is a useful complement to conventional CAD work for a number of tasks, such as: digital pre-assembly; planning analysis; manufacturing instructions *in situ*; installation design (hydraulics, mechanical, …); and selection of parts to be extracted from the digital mock-up. Also non-immersive VE is a good environment for collaborative design through the use of large stereoscopic screens and polarizing glasses. Moreover, the combination of video-conferencing technology with stereoscopic graphics displays has been reported elsewhere (Potter, 1995).

Agents (Wooldridge and Jennings, 1994) represent a promising technology to be used by Virtual Environments in CAD systems. Interface agents can be envisaged as assistants floating in Virtual Environments linked to the intranet. In this context, these agents can spread out the user's intentions over the intranet. This subject, however, is still highly speculative and belongs to the agenda research of the academic laboratories. An interesting discussion on agents and solid modeling systems can be found in Feijó et al. (1996).

CONCLUSIONS

Irrespective to the built-in degree of Artificial Intelligence, any tool for assisting structural and mechanics engineering design is deeply rooted into 3D environments. However, the question of 3D virtual worlds is not clearly presented in the AI-based CAD literature.

This paper contributes to the understanding of Virtual Environments in CAD systems in a number of ways. Firstly, a more adequate definition of VE, as a radial category, based on a more formal structure is proposed. Secondly, virtual workspaces (VW) in Non-immersive VE are classified in terms of the fixation point (Front and Rear VW). Thirdly, the most relevant VE technologies for CAD systems are presented. Finally, the question of

VE assisting the design process is discussed and the link of virtual objects with design views is established. Amongst the future research issues, this paper recommends the use of intelligent interface agents in Virtual Environments for CAD systems.

This paper is also a practical guide for CAD researchers and engineers who want to explore the possibilities of VE technology. VE technology is still in its infancy but the benefits from it have been widely recognized. The best decision for the use of VE in integrated CAD system is to focus on products that have been used by most of the companies around the world. In this regard, Immersive VE technology for CAD environments is well represented by BOOM-type devices. From the non-immersive side of VE, the following technologies seem to be adequate for CAD environments: (1) stereo glasses/monitors; (2) Cyberscope-like technologies; (3) large stereoscopic screens for collaborative engineering; (3) PHANToM-like haptic interfaces; (4) Responsive-Workbench-like technologies; (5) avatar applications using 6 degree-of-freedom position/orientation sensors, stereo glasses and gloves. The possibility of using Augmented Reality in an integrated CAD system should be restricted to tasks that do not require accuracy. In this context, typical usage of Augmented Reality are the following: documentation handling; inspection instructions; and manual manufacturing processes (see Sims (1994)).

ACKNOWLEDGMENTS

The author would like to thank the CNPq for financial support and Prof. João Bento for valuable discussions on CAD matters.

REFERENCES

Badler,N.I.; Phillips,C.B. and Webber,B.L. 1993. Simulating Humans - Computer Graphics Animation and Control, Oxford University Press, New York, NY.

Bolas,M.T. 1994. Human factors in the design of an immersive display. *IEEE CG&A*, January 94, p. 55-59.

Bryson,S. 1996. Virtual reality in scientific visualization. *Comm. ACM*, v.39, no.5, May 96, p. 62-71.

Bryson,S. and Levit,C. 1992. The virtual wind tunnel. *IEEE CG&A*, July 92, p. 25-34.

Cruickshank,D. 1996. Information is a place. *IRIS universe*, v.36, Summer 96, p. 17-21.

Cruz-Neira,C.; Sandin,D.J. and DeFanti,T.A. 1993 Surround-screen projection-based virtual reality: the design and implementation of the CAVE. Proc. SIGGRAPH 93, ACM Press, New York, p.135-142.

Deering,F.D. 1996. The HoloSketch - VR sketching system. *Comm. ACM*, v.39, No.5, May 96, p. 54-61.

Ellis,G. 1996. Digital clay: transforming automobile design. *IRIS Universe*, no. 37, Fall 96, p. 28-32.

Feijó,B. 1992. Perception and Cognition in Intelligent CAD Systems. In C. Tasso and E. Arantes e Oliveira (eds.), Expert Systems in Structural and Mechanics Engineering, Course Notes, Advanced School/CISM, 6-12 July, Udine, 1992.

Feijó,B.; Lehtola,N.; Bento,J. and Scheer,S. 1996. Reactive design agents in solid modelling. In J.S.Gero and F.Sudweeks (eds.), Artificial Intelligence in Design '96, Kluwer Academic Publ., p. 61-75.

Feiner,S.; MacIntyre,B. and Seligmann,D. 1993. Knowledge-based augmented reality, *CACM*, v.36, no.7, July 93, p. 53-62.

Glassner,A.S. 1995. Principles of Digital Image Synthesis, Morgan Kaufmann, San Francisco, CA.

Goel, V. and P. Pirolli. 1989. Motivating the notion of generic design within information-processing theory: the design problem space, *AI Magazine*, Winter (1989), p.19-36.

Krueger,M.W. 1991. Artificial Reality II, Addison-Wesley, Reading, Mass.

Krueger,W. and Froehlich,B. 1994. The responsive workbench, *IEEE CG&A*, vol.14,No.3, May 94, p. 12-15.

Lakoff,G. 1987. Women, Fire, and Dangerous Things: What Categories Reveal about the Mind. University of Chicago Press, Chicago.

Marcus,A. 1993. Human communications issues in advanced UI's, *Comm. ACM*, Vol.36,No.4,April 93, 101-109.

McKenna,M. and Zeltzer,D. 1992. Three dimensional visual display systems for virtual environments, *Presence: Teleoperators and Virtual Environments*, v.1, no.4, p. 421-458.

Nielsen,J. 1993. Noncommand user interfaces, *Comm. ACM*, v.36, no.4, April 93, p. 83-99.

Potter,C.D. 1995. Digital tools for collaborative engineering, *Computer Graphics World*, v. 18, no. 8, August 95, p. 51-59.

Regan,M. and Pose,R. 1994. Priority rendering with a virtual reality address recalculation pipeline, Proc. of SIGGRAPH 94, Computer Graphics Proc., Annual Conference Series, ACM SIGGRAPH, p. 155-162.

Sims,D. 1994. New realities in aircraft design and manufacture. *IEEE CG&A*, v.14, no.2, March 94, p. 91.

Takala,T. and Hahn,J. 1992. Sound rendering, *Comp. Graphics*, Proc. SIGGRAPH'92, ACM, Vol. 26, No. 2, July 92, p. 211-220.

Tasso and Arantes e Oliveira,E. 1992. Expert Systems in Structural and Mechanics Engineering, Course Notes, Advanced School/CISM, 6-12 July, Udine, 1992.

Vince,J. 1995. Virtual Reality Systems, SIGGRAPH Series Book, ACM Press.

Wann,J. and Mon-Williams,M. 1996. What does virtual reality NEED?: human factors issues in the design of three-dimensional computer environments, *Int. J. Human-Computer Studies*, v.44, p. 829-847.

Wooldridge,M.J. and Jennings,N.R. 1994. Agent theories, architectures, and languages: a survey, *Proc. ECAI94 Workshop on Agent Theories, Architectures and Languages*, Amsterdam, The Netherlands, p. 1-32.

MODEL TRANSMUTATIONS FOR CONCEPTUAL DESIGN
OF TECHNICAL SYSTEMS

E. Toppano
University of Udine, Udine, Italy

Abstract. In this paper, engineering design is considered from the point of view of modeling i.e. the construction and manipulation of models of (possible) physical realities. Consequently, the design activity has been analysed in terms of patterns of inference called model transmutations. Three categories of transmutations namely, transformations, combinations and retrievals, have been discussed with reference to the Multimodeling approach for representing physical systems. The major goal of the paper is to provide a conceptual framework for analysing existing design systems and for addressing questions concerning their competence such as what types of inference patterns underlie different design strategies e.g. top-down, compositional and analogical design; what kind of design solutions a design system is able to generate from what kind of input specification and prior design knowledge; what is the logical relationship between specification and prior design knowledge. A second goal is to provide a basis for the development of a general theory for task adaptive multistrategy design that aims at combining a range of different design strategies dynamically, in order to take advantage of their respective strengths and address a wider range of practical problems.

Key words: design process, multistrategy design, model transmutations

1. Introduction

Engineering design can be abstractly characterized as a constrained function-to-structure mapping [7], [8], [18]. It takes as input a functional specification of the artifact to be built, including desired goals and constraints on design, and a description of the available technology and of general physical principles. It produces as output a description of an artifact that satisfies the specification and contains enough information to allow the manufacturing, fabrication or construction of the desired system. One method for solving design problems is PCM i.e. propose, critique, and modify [4]. The method have the subtasks of proposing partial or complete design solutions, verifying proposed solutions by identifying causes of failure if any, and modifying proposals to satisfy design goals.

In this paper, we characterize the PCM method from the point of view of modeling i.e. the construction and manipulation of models of (possible) physical realities. Under this perspective the design process consists of a sequence of cycles. In each cycle the designer analyses the current design solution (i.e. a possibly incomplete model of the desired artifact) in terms of his/her background knowledge and problem specification and decides which action (i.e. model manipulation) to do next in order to improve the solution. In contrast to decision making where alternative generation is usually completed before evaluation is begun, in design there is a close interaction between these activities. Model evaluation supports discovery of new goals and objectives (i.e. respecification), the knowledge of which informs the subsequent generative activities.

One goal of the paper is to identify what kind of inferences, thereafter called model transmutations, underlie different design processes such as top-down, compositional and analogical design. We will focus on the early phase of design that is, conceptual design, which results into the topology of the desired artifact without stating definite values for all constructive parameters (e.g., geometrical, physical, etc.). The analysis is aimed at providing a conceptual framework by means of which: i) the design systems proposed in literature can be studied and compared on the base of the inference capabilities they presuppose i.e. on the base of their levels of competence; ii) model transmutations can be used as building blocks to explain or experiment with different design strategies. A second goal is to identify what is the logical relationship among model transmutations, input information (i.e. design specification) and the designer's prior knowledge in order to understand the preconditions and the circumstances under which each transmutation can be used. The presented ideas provide a basis for the development of a general theory for task-adaptive multistrategy design. By multistrategy design we intend the composition of two or more types of inferences in the same design process. The composition is made dynamically as the design process unfold according to the demands of the current situation.

The paper is organized as follows. In section 2 we briefly survey the main concepts of the Multimodeling approach for representing and reasoning about physical systems which constitutes the background of the present work. Advantages of using the multimodeling approach for representing design knowledge are discussed in section 3. We then introduce four basic elements of design problems - namely design specification, operational model, background knowledge and design operator - which are used to model the design process (sections 4 and 5). The next two sections are devoted to illustrate elementary transmutations and to show some examples of multitype inferences using transmutations. In section 8 we discuss relationships between design problems and background knowledge. Finally, section 9 illustrates the main features of SECS, a system developed to experiment with multistrategy design in the electrical domain, while section 10 discusses related work and draws conclusions.

2. Background: the Multimodeling approach

In recent years, a novel approach to the representation and reasoning about physical systems, called *multimodeling*, has been proposed. This approach is based on the key idea of considering the task of reasoning about a physical system as a co-operative activity which exploits the contribution of many diverse models (i.e. knowledge sources) of the system at hand each one encompassing a specific type of knowledge and representation. The execution of a problem solving task (e.g. supervision, diagnosis) within the multimodeling approach is based on two fundamental mechanisms: (1) reasoning *inside* a model, which exploits knowledge available within a single model by using model specific problem solving methods, and (2) reasoning *through* models, which supports opportunistic navigation among models in order to allow each individual step of the problem solving activity to exploit the most appropriate knowledge source. A detailed description of representation and reasoning issues in the multimodeling approach is given in [5]. For the purpose of this paper we are mainly concerned with knowledge modeling and organization.

2.1 The concept of model
We assume here that a model is basically a device which is built to answer specific questions about some portion of reality. In particular, given two objects M and S and an observer O, the object M is said to be a *model* of the object S if the observer O can use it to answer questions Q - or, more generally, to perform tasks - that interest him about the object S. Two important consequences of the above definition are that:
- a model is a "surrogate", a substitute for the real system, that is developed to enable the observer-user to determine consequences by reasoning about the world rather than taking actions in it. Being a surrogate, it is unavoidably an abstraction of reality and embodies a host of implicit assumptions which are specific to its intended purpose;
- for an object under investigation there is not "the model" to represent it but a set of models representing it from different perspectives, at different levels of abstraction and for different purposes according to observer O, type of question (or task) Q and application objectives.

Though model specificity is unavoidable it is not undesirable because it often buys us computational efficiency in reasoning. Moreover, for a variety of complex problem solving tasks it is difficult to identify in advance the most appropriate representation. As a consequence, the model builder must provide the reasoner (human or artificial) with several different models of the real system to be integrated dynamically in the course of solving a particular problem.

2.2 Modeling assumptions
In the multimodeling approach the specification of a model includes the intended purpose of the model together with the modeling assumptions lying behind its construction. Modeling assumptions have been divided into four main categories: ontological commitments, epistemological types, operating assumptions, and representational assumptions. We briefly illustrate these categories below.
- *Ontological commitments* concern the types of entities, relations and properties that are presupposed to exist in the real world and thus can be represented in the model. For example, an engineer may think of the liquid in a container as an individual object characterized by macroscopic properties such as its temperature and pressure or consider the same liquid as a population of molecules characterized by their positions, velocities and kinetic energies. Several ontologies have been proposed to represent physical systems. See [2] for a thorough discussion of various alternatives.
- *Epistemological types* refer to the type of knowledge that a model can represent about reality. We identify five epistemological types:
  Structural knowledge i.e. knowledge about system topology. This type of knowledge describes which components constitute the system and how they are connected to each other.
  Behavioral knowledge i.e. knowledge about the potential behavior of components. This type of knowledge describes how components can work and interact in terms of the physical quantities that characterize their state and the physical laws that rule their operation;
  Functional knowledge i.e., knowledge about the roles components may play in the physical processes in which they take part. This type of knowledge relates the behavior of the system to its goals and deals with functional roles, processes and phenomena.
  Teleological knowledge i.e., knowledge about the goals assigned to the system by its designer, the expected behavior of the system and the operational conditions that allow the achievement of the goals through correct operation;

<u>Empirical</u> knowledge i.e. knowledge concerning the explicit representation of system properties through direct empirical associations. This type of knowledge may be derived from observation, experimentation, and direct experience with the system.
Ontological and epistemological assumptions provide the majority of the vocabulary for representing the system under consideration.

* *Operating assumptions* concern the applicability of the model and indicate its correct moment of use. They specify the conditions (initial conditions or region boundary conditions) that must be satisfied by the real system in order for the model to be considered as a valid representation of reality and to be used to achieve its purpose. For example, modeling the behavior of a bipolar junction transistor in the forward-active region requires a specific bias condition, that is, forward bias the emitter junction (Veb >0) and reverse bias the collector junction (Vcb <0).

* *Representational assumptions* refer to the coverage, detail and resolution of the model. *Coverage* specifies the range of phenomena that are considered relevant to the purpose of the model and, thus, must be explicitly included in the model and the kind of simplifying assumptions that are appropriate. For example, a wire can be modelled as an electrical conductor by ignoring inductive phenomena or it can be represented as an inductor that explicitly models the magnetic field generated by the current flowing through it. Moreover, the conductor may be modelled as a perfect conductor, a constant resistor, a temperature dependent resistor and so forth. *Detail* specifies the degree of granularity of the represented knowledge. For example, the structural model of a system can be represented at the level of major subsystems or can be further refined at the level of elementary components. Finally, *resolution* specifies the degree of precision of the results attainable by reasoning with the model. For example, resolution can be lowered by relaxing real valued variables and using qualitative (behavioral) models.

The explicit representation of modeling assumptions within a model facilitates model selection and reuse and provides control during problem solving so that only the relevant knowledge is used.

2.3 Model organization

In the multimodeling approach any choice about ontology, is allowed, as well as any kind of epistemological type, representational and operating assumptions. The only restrictions that are imposed to the organization of the various models of a system are the following:

- the model base is strictly layered i.e. any individual model is allowed to encompass only one specific choice about ontology, epistemological type, and representational assumptions. For example, we never mix concepts belonging to different ontologies or epistemological types in the same model. One advantage of this choice, in addition to modularity, is the ease of controlling multilevel reasoning;

- models are not independent, i.e. any individual model is based on the existence and on the characteristics of other models. As a consequence models must be explicitly and appropriately connected to each other by links. *Links* are bidirectional relations that connect corresponding knowledge elements in different models. For example, the link between a structural model and a behavioral model of a system describes which physical quantities and physical equations in the behavioral model are associated to which terminals, components and nodes in the structural model. The link between function and behaviour is established by associating equations governing the behaviour of components in the behavioural description with appropriate functional roles (e.g., generator, conduit, barrier, reservoir) in the functional (role) model. Note that, in general, a component may play different functional roles in the same or in different physical domains. An electrical resistor, for example, is a conduit of electricity in the electrical domain and a generator of heat flow rate in the thermal domain. Links within the functional description associate i) cofunctions - namely, functional role networks capable of supporting primitive processes such as transporting, reservoir charging and discharging - to the processes they describe, and ii) organisations (i.e. process networks) to the phenomena they support. Finally, the link between function and teleology is realised by associating goals in the teleological description with the phenomena (or the primitive processes) represented in the functional model which are used to achieve them. The relation between components and goals is many-to-many since a component may participate to the realisation of several goals and, conversely, a goal can be fulfilled by utilising the behaviours of several components of a device.

Since models may employ different ontologies a link may also specify a set of bridge rules (i.e., rules whose premises and conclusions belong to different models) which allow the exportation of partial results from a model to the other. Thus, links have two roles: i) they allow to shift the focus of attention from part of a model to a related one in another model and ii) they support a very simple form of knowledge translation between representations.

Technically, the representation of a system in the multimodeling approach is specified by a graph R=(M, L) where: M={Mi} is the set of models of the real system and L={Lij} is the set of links between models in M. Each model Mi is described through three components: i) a system description i.e. a set of facts about the system at hand expressed in some representation language (e.g. first order logic), ii) a set of reasoning utilities that can be used with the system description to perform a set of model specific tasks called *elementary tasks* and iii) a model context representing the intended purpose of the model together with the main assumptions lying behind its construction. A characteristic feature of the multimodeling approach that makes it different from other approaches that use multiple representations such as FBS [23] and the FR scheme [8] is that substantive reasoning can occur at any level i.e. the higher levels (functional, teleological) are not merely for control and explanation but can be used by themselves to perform complex tasks (e.g. diagnosis). We call this property: multilevel operationality.

3. Using the multimodeling approach in design tasks

When dealing with design one is invariably concerned with representation systems. The choice of a representation system strongly influences the design processes and its products. In particular, the representation used determines the type of knowledge about an artifact and the constraints designers are able to consider. The multimodeling approach for representing physical systems has several noteworthy characteristics that can be usefully exploited in design tasks.

First, the concept of model used in the multimodeling approach supports the explicit representation of a model contextual information such as the ontological, representational and operating assumptions lying behind its construction. As a consequence models developed using this approach may serve as a record of commitments, helping the designer i) to focus only on relevant information at each stage of the design process, ii) to maintain system coherence among representations or iii) to represent mutually incoherent description of the same artifact (e.g. the description of an artifact's behavior under different operating modes).

Second, the approach supports the representation of several types of knowledge about an artifact such as structural, behavioral, functional and teleological knowledge. Because design specification is usually expressed in a language remote from solution description, this feature is used to support the conceptual mapping as from goals, into function, then into the artifact's behavior, and, finally, into its structure that design effects. Moreover, the use of separate descriptions each one devoted to an epistemological type allows for a better focusing of the design process since it is possible to take into account only those pieces of knowledge that are relevant in a given stage of the design process (e.g., teleological and functional descriptions in the early phase of design, behavioural descriptions in the phase of analysis and evaluation of design solutions). As a consequence, efficiency and cognitive plausibility are improved;

Third, the possibility to representing the designed system at different levels of detail and accuracy (i.e. coverage and resolution) supports incrementalism i.e., a step-wise production of descriptions of lower and lower aggregation level and of greater accuracy. This enables the designer to exploit a step-by-step methodology, thus reducing the computational complexity of the overall design process.

Fourth, by explicitly representing links between knowledge elements in different representations, the approach support the co-operative use of models for explanation and evaluation. Following the links bottom-up (i.e. from structural models to teleological models through behavioral and functional ones) it is possible to provide a teleological explanation of the behavior of a component or to explain the reasons behind a particular arrangement of components. This is valuable when we want to evaluate the effect of a modification of a design such as the elimination of a component or the substitution of a component with another one. Analogously, it is possible to follow the links top-down (i.e. from teleological models to structural ones) to assign blame to components i.e. to identify which components are responsible for the achievement of a given set of goals. This task is accomplished, for example, when it is needed to establish which part of a structural description can be reused to fulfil a given set of new goals. Finally, links enable quantified estimation of several metrics such as those proposed by Kannapan [11]. This kind of evaluation is impossible if the relations between teleology, function, behaviour and structure are ignored or are specified only in "ad hoc" manner.

4. Elements of design problems

Our analysis of design problems is constructed around four conceptual elements: design specification, operational model, background knowledge, and design operator. These elements are briefly described below.

Design specification (SPE) refers to design goals and constraints. *Design goals* represent the ultimate purpose that the design is intended to achieve when put in use. *Constraints* express relations among properties of the proposed artifact (or the process by which it is designed or manufactured) and its environment. They may refer, for example, to physical characteristics of the artifact (e.g. its size, weight, structure, appearance), to performance characteristics (e.g. operational efficiency and reliability), to resource (e.g. time, money, materials, and expertise) availability, etc.

The *operational model* (M) is a, possible incomplete, representation of the artifact to be designed. It is specified by one or more design fragments. According to the multimodeling approach a *design fragment* (DF) is constituted by a system description and a context. The system description is a set of statements describing some properties of the artifact under consideration. The context specifies the ontology (i.e. the types of entities, relations, and attributes with associated domains of values) used to build the system description as well as a set of operating and representational assumptions that explicitly state conditions for its validity. The design fragments constituting the operational model may differ in several ways. Some of them, for example, may represent different components or subsystems of the artifact to be designed. Other fragments may represent the same component using different ontologies or epistemological types (e.g. a fragment describes a component structure while another fragment describes its behavior); yet other design fragments may describe the same component using different operating and representational assumptions.

The design specification and the operational model coevolve during the design process: goals and constraints are gradually turned into descriptive statements that are included into the operational model; at the same time design analysis and evaluation supports the discovery of new constraints and objectives which modify design specification. More specifically, let $T=\{0, 1, \ldots n,\ldots\}$ be a discrete set of time points. By SPE(n) and M(n) we denote the specification and the operational model at time point n. We then define the *design state* at time point n, thereafter DS(n), to be the pair [SPE(n), M(n)].

Background knowledge (BK) represents the prior designer's knowledge. This knowledge includes two main components: physical knowledge and design knowledge. *Physical knowledge* represents the understanding that engineers have of physical phenomena. It consists, for example, of general principles and laws of physics, techniques of applied mathematics, ontologies of engineering concepts. *Design knowledge* refers to empirical knowledge that describes how design can be performed. This knowledge may include, for example, libraries of design exemplars such as design cases or design prototypes (i.e. generalizations of cases) in various domains of application, knowledge about the possible decompositions of design goals into subgoals, means-ends knowledge (e.g. knowledge about how goals can be attained by the use of functional compositions together with the behaviors and structures supporting these functions), general principles of good design that may be used to guide the design process. The content and organization of the background knowledge strongly influence the design process and its results as it will be shown in the following sections where several examples of physical and design knowledge are illustrated.

Design operators (0) represent actions that can be performed during the design process. We distinguish two categories of design operators:
- *domain operators* i.e. operators that change the design state. They are organized into three groups depending on whether they
    1) transform existing elements of the design state (e.g. a design fragment or a constraint) according to some rule (Transformations);
    2) combine together two or more existing elements of the design state (Combinations);
    3) retrieve from background knowledge new elements (e.g. a case or prototype fragment) that are added to those already existing in the current design state (Retrievals).
  Collectively, these abstract operators are called *knowledge transmutations*. Seventeen specific operators have been identified as specializations of them. Section 6 discusses these operators in detail.
- *decision operators* i.e. operators for evaluating design states and for selecting and scheduling domain operators. This category includes:
    1) simulation operators for predicting actual behavior (or function) of a proposed operational model;
    2) comparison operators for detecting discrepancies between the properties of the actual (or predicted) design state and design specification;
    3) diagnostic and repair actions for finding causes of discrepancies and propose remedies (e.g. create new goals or objectives to fix a bug in the design solution);
    4) selection operators for choosing among a specified set of alternatives according to some criterion.

No claim is advanced that the above list is complete or that the partitioning of functionality is ideally chosen. Rather, this description is offered as an example of the types of operations that are observed in design processes.

5. The design process

Observation of human designers while trying to solve a design problem, indicates that they do not follow a rigid plan detailing what action to do next. Rather, they move fluidly between various problem pieces (e.g., design fragments) and design actions (e.g., from generation, to evaluation of solutions and problem reformulation) in a flexible and highly opportunistic way [25]. Being able to take advantage of opportunities requires being able to judge whether progress has been made along a certain line of attack, and to choose which ideas are more promising or more likely to lead to novel solutions. This observation suggests that much of the thought process is actually at the meta-level, that is, it is *about* the process of designing an artifact. According to this view, we adopt a two layered structure to model the design process with a separate problem solver for the meta level. More specifically, we assume that the design process is constituted at the domain level (or base level) by a sequence of modeling steps which are guided at the meta level (or control level) by a decision activity which continuously monitors and redirects the activity at the domain level. *Modeling steps* describe transitions between design states. A modeling step takes as input: the current design state $DS(n)$, a knowledge transmutation operator $TR(n)$ and background knowledge BK. It produces as output a new design state $DS(n+1)$ obtained by applying the specified transmutation to the current design state. At the control level, the decision activity can be decomposed into a sequence of decision steps. Each *decision step* takes as input the current design state $DS(n)$ and background knowledge BK. It produces as output a transmutation operator $TR(n)$ and (possibly) a modified design state $DS^*(n)$ (e.g. additional constraints). The decision activity: i) analyses and evaluates the actual design state, ii) identifies what it needs in order to improve the overall design activity, iii) formulates internal objectives that are included in design specification, and iv) select the knowledge transmutation that is most appropriate for pursuing these objectives. Hence, in our conceptualisation, the design process results in a ordered sequence of design states $<[SPE(0), M(0)], ..., [SPE(k), M(k)]>$, with $SPE(0)$, $M(0)$ representing the initial specification and operational model, respectively, and $M(k)$ representing the final design solution i.e. a complete and sufficiently detailed operational model of the desired artifact that satisfies specification $SPE(k)$. In order to cope with the complexity of design problems the above process is usually decomposed into several phases at different level of abstraction. For example, conceptual design, embodiment design and detail (or concrete) design are three different abstraction levels of the design process which are commonly applied in mechanical engineering design [18]. In this paper we will focus on conceptual design. The main objective of this phase is to synthesise a primitive solution concept of a new engineering device or to improve an existing one in an innovative way (e.g. by using already known working principles in an unusual way). We decided to focus on conceptual design because it is generally the more difficult and innovative phase of the design process and because decisions taken in this stage have a major impact on what follows.

6. Model transmutations

Model transmutations describe the basic inferences that can be performed for constructing, modifying or combining design fragments using background knowledge. We use the term inference in a generic sense to mean any way to get new information from old. In this section we discuss some important types of transmutations, but the list is hardly complete.

Model transmutations have been classified into three classes: transformations, combinations and memory retrievals. Although the discussion focuses on operational models, most of the following transmutations - such as generalisation, specialization, conceptual abstraction and concretion - can be performed also to modify design specification.

6.1 Transformations

Transformations are unary operations that take as input a design fragment DF and produce as output a new design fragment DF* by modifying the elements of DF according to some rule. The application of a transformation usually requires specific background knowledge. According to [13] we assume that this knowledge is organized in one or more concept hierarchies. A *concept hierarchy* is composed by nodes representing abstract or physical entities, and oriented arcs representing frequently occurring relationships among the entities such as "type-of", "part_component", or "precedes". Transformations can be classified according to two main criteria: 1) the type of concept hierarchy used by the transformation and the direction (up or down) followed by the operation along the hierarchy, and 2) the target element of the transformation, that is, the specific aspect of a design fragment (the ontology or the system description) that is affected by the transformation. A brief description of basic transformations follows.

• *Aggregation* (versus *refinement*): different sets of descriptive statements are grouped into a single set following a "part-component" hierarchy. According to the ontology of the design fragments that are involved in the operation it is possible to further distinguish among structural, behavioral, functional, and teleological aggregations (refinements). Figure 1 (fragments DF0, DF1) shows an example of behavioral aggregation (refinement). A set of physical equations (c1-c5) representing the behavior of a simple RC circuit in the source fragment DF0 is contracted into a single equation (c1) in the transformed fragment DF1.

• *Approximation* (versus *elaboration*): descriptive statements are replaced by other statements that are less accurate according to some precedence hierarchy but closely resemble the original ones. There are several types of precedence hierarchies. One type is the rank hierarchy. It consists of values representing the "rank" of an entity in some structure. For example, equations of physics that represent the same aspect of reality based on different and contradictory assumptions can be organized into rank hierarchies according to a "simpler_than" relationship. The rank hierarchy RH1 shown in Figure 2 organises alternative ways to model the voltage across a battery. In Figure 1 (fragments DF0, DF2) it is shown a behavioral approximation based on this type of rank hierarchy. The operation is performed by replacing the physical equation $V1=E=f(\underline{CL})$ in DF2 with the "simpler" equation $V1=\underline{E}$ in DF0. (We underline a variable to mean that it is constant).

• *Reduction* (versus *expansion*): descriptive statements representing a specific aspect of reality or a class of phenomena are removed from a set of descriptive statements. As an example, design fragment DF0 in Figure 1 is obtained by removing from the source fragment DF3 the equation $B=kI/r$ representing the phenomenon of magnetic induction of electrical current. Note that the reduction operator (and its inverse) can be used to control which aspects of reality are considered immaterial for the problem at hand and thus must be eliminated from the representation and which aspects, on the other hand, are considered relevant and thus must be included. Instead, the approximation operator (and its inverse) can be used to control how relevant phenomena must be represented i.e. what kind of simplifying assumptions are appropriate. The application of these transformations

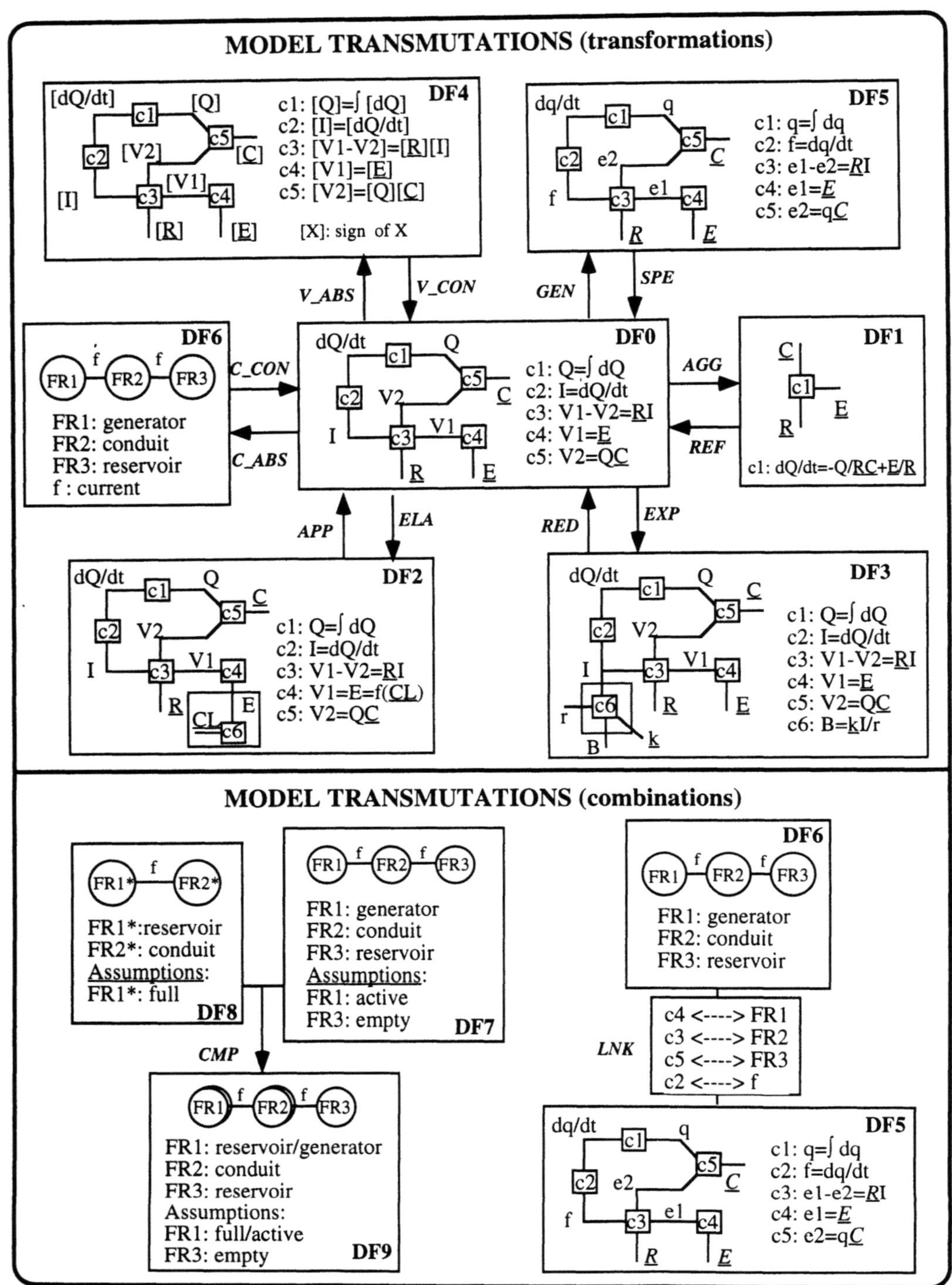

Figure 1 Examples of model transmutations

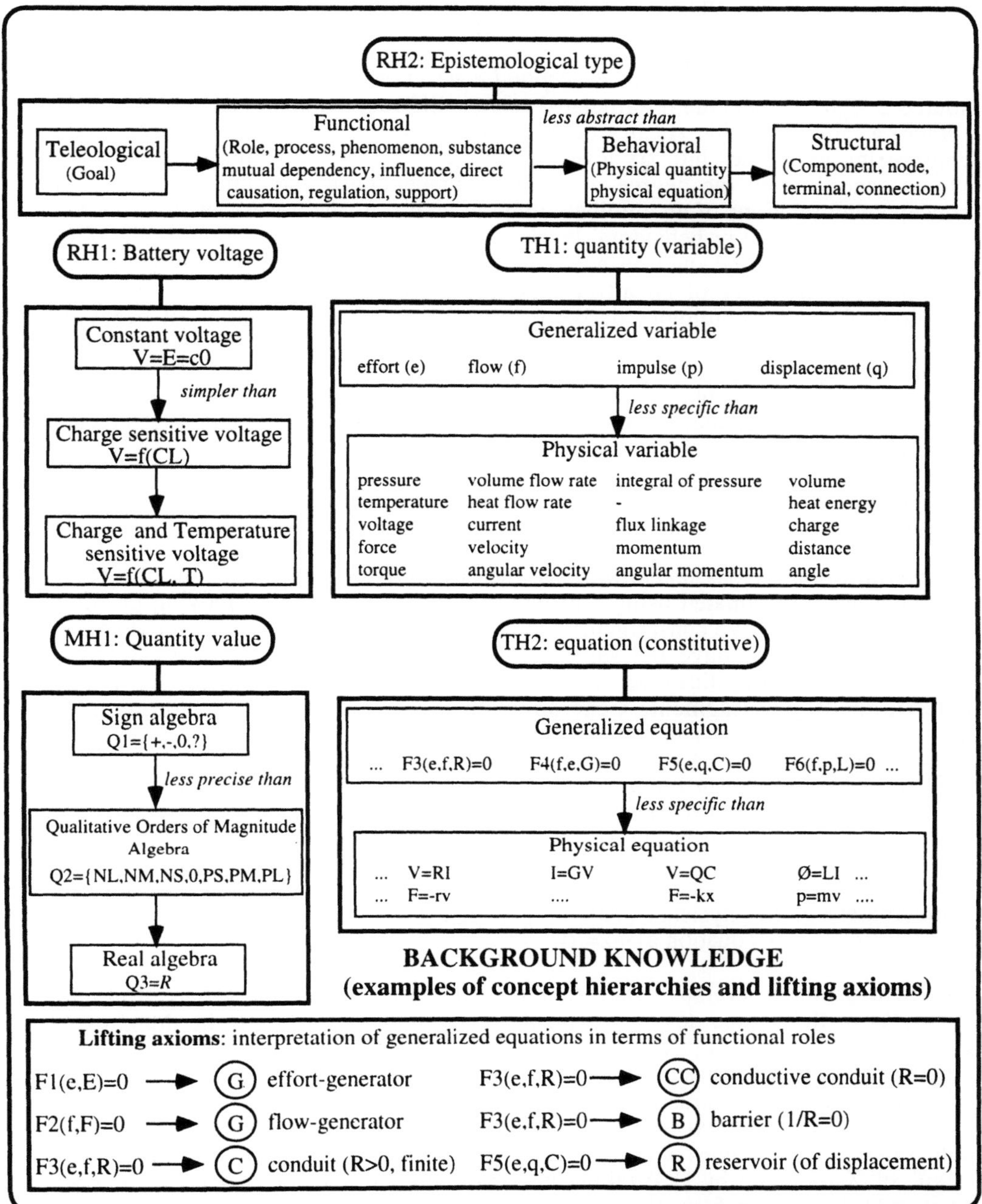

Figure 2 Background knowledge (partial view)

produces, as a side effect, the modification of the context (i.e. the representational assumptions) of the resulting design fragments.

• *Conceptual abstraction* (versus *conceptual concretion*): the descriptive statements of the source design fragment are reformulated using a different ontology obtained by moving up (down) a rank hierarchy. As an example, the rank hierarchy RH2 in Figure 2 organises engineering ontologies into six levels of abstraction (corresponding to epistemological types) according to the degree of context independence and function neutrality the concepts and relations of an ontology allow to attain. At the lowest levels of abstraction (e.g. at the structural or behavioral levels) concepts and relations represent the objective language of physics and engineering sciences which is relatively neutral with respect to the intended functionality of an artifact. At the higher levels (e.g. at the functional or teleological levels) ontologies derive from a subjective interpretation of behavior and structure in a context. Conceptual translations between levels of the hierarchy require a set of *lifting axioms* specifying how the concepts and relations of an ontology are appropriately reinterpreted in the light of a new ontology. Figure 1 shows an example of conceptual abstraction (concretion). The behavioral design fragment DF0 is abstracted into the fragment DF6 by reformulating physical quantities and equations in the source fragment in terms of a functional role network i.e. a series of three functional roles namely a generator, a conduit and a reservoir connected by mutual dependency relations. The transformation is based on the lifting axioms shown in Figure 2 (bottom).

• *Generalization* (versus *specialization*): the descriptive statements of the source design fragment are reformulated using a different ontology obtained by moving up (down) a "type-of" hierarchy. We constrain the two ontologies to share the same level of conceptual abstraction i.e. they have the same epistemological type. As an example, Figure 1 (fragments DF0, DF5) illustrates an example of generalization (specialization) involving a behavioral design fragment. Generalization is performed by substituting physical quantities and equations in the behavioral design fragment DF0 with corresponding generalized entities. The transformation is based on the use of the typological hierarchies TH1, and TH2 shown in Figure 2. According to hierarchy TH1, physical quantities Q (electrical charge), I (current), V (voltage), R (resistance), E (electromotive force) and C (capacity) are replaced by corresponding generalized quantities namely, q (displacement), f (flow), e (effort), R (generalized resistance), E (generalized electromotive force) and C (generalized capacity). Similarly, physical equations are replaced by their generalized counterparts e.g. the physical equation V1-V2=$\underline{R}$I (i.e. an instance of the the Ohm's law) is generalized into the equation e1-e2=Rf.

• *Value abstraction* (versus *concretion*): small differences in the values of descriptive attributes in the source design fragment are ignored by replacing the domains of the attributes (i.e. the sets of possible values for the attributes) with new domains obtained by moving up (down) a measure hierarchy [14]. As an example, Figure 1 (fragments DF0, DF4) shows a value abstraction (concretion) applied to the attribute "value" of physical quantities in a behavioral design fragment. Value abstraction occurs by replacing the numerical (i.e. real) values of each quantity X in the source design fragment DF0, with the "sign" of X (denoted by [X]) taking possible values in the set {negative (-), zero (0), positive (+)}. The transformation is based on the rank hierarchy MH1 shown in Figure 2.

6.2 Combinations
Combinations are binary operations that take as input two design fragments DFi, DFj and produce as output a new design fragment DF* by combining the elements of DFi and DFj

according to some specified rule. We identified two types of combinations: compositions and links.

• *Composition* (versus *decomposition*): two design fragments having the same ontology are composed by forcing two previously separated elements of their system descriptions to be the same, thereby, requiring to add a codesignation assumption in the output design fragment. The rule associated to the transmutation specifies compatibility conditions that the elements selected to be unified must satisfy. The operational and representational assumptions associated to the output design fragment are the union of the respective assumptions in the source design fragments. As an example, Figure 1 (bottom left) shows the composition of two functional (roles) fragments (DF7 and DF8). The fragments are composed by unifying compatible elements. Two roles are compatible if they are equivalent roles and i) they are identically instantiated, or ii) they are not yet instantiated or iii) one role is instantiated and the other is not. In the above example FR2* and FR2 are unified since they represent identical roles (i.e. two conduits of the same flow f). Analogously, FR1* (a reservoir) and FR1 (a generator) are unified according to a rule that establishes the equivalence of a not empty reservoir and an active generator. Notice that FR1* and FR3 (two reservoirs) are not compatible since they make contradictory operating assumptions (i.e. FR1* is assumed to be full while FR3 is empty).

• *Link* (versus *unlink*): the elements of the system description of a design fragment DFi are put in correspondence with the elements of the system description of another design fragment DFj. The set of entities described or referred to by the two fragments cannot be disjoined. Fragments DFi and DFj may differ in ontology (e.g. they represent different levels in the epistemological hierarchy), in detail (e.g. DFi and DFj have different aggregation level) or resolution (e.g. they represent different levels in a measure hierarchy). Figure 1 (bottom right) shows an example of link between a behavioral (DF5) and a functional (DF6) design fragment. The operation is realised by associating equations governing the behavior of components in the behavioral design fragment with appropriate functional roles in the functional fragment.

6.3 Retrievals

Retrievals are transmutations that generate a design fragment from scratch by recalling a design exemplar (a case or a prototype) from design background knowledge. We assume that design exemplars are represented in the same way as operational models that is, by a collection of interlinked design fragments each one devoted to describe a specific class of properties (e.g. structural, behavioral, functional, and teleological) about the artifact represented by the exemplar. The operator takes as input a library of design exemplars, a set of indexes (i.e. features of the input problem that are deemed relevant to finding similar exemplars), and a similarity metric. The operator matches design exemplars against the input features and uses the similarity metric to retrieve those exemplars that are most similar to the input. To this end the design exemplars must be labelled and organized so that features of input problems can be used to find them.

7. Multitype inference in conceptual design

Basic transmutations can be concatenated in various ways to produce complex inference patterns. Some recurring combinations of operators have a particular significance. For example, the operation of *similarization* (also known as cross contextual analogical reasoning) substitutes the ontology of a design fragment with a sibling ontology in a typological hierarchy and reinterprets the system description of the source fragment in the new conceptualisation. This operation can be viewed as the concatenation of a generalization transformation followed by a (re)specialization in a different physical domain.

As an instance, suppose that the problem is to design an aircraft rate-of climb sensor (denoted by Xb) whose purpose (G*) is to measure the rate of pressure change of a given pressure source. Suppose further that no prior exemplar exists in design background knowledge for this class of devices. The problem may be addressed by exploiting cross contextual analogical reasoning as follows. First, the desired goal "G*: TO_SENSE_RATE_of p: pressure CHANGE" is generalised into "G: TO_SENSE_RATE_of e: effort CHANGE" by recognising that pressure is an instance of the generalised effort variable "e" in the hydraulic domain (see the typological hierarchy TH1 in Figure 2). Second, the library of design exemplars is searched to find an exemplar whose teleological representation describes a goal that is a specialization of G in some physical domain. Suppose that such a specialisation exists in the electrical domain so the search succeeds giving an exemplar (say P1) for the class of electrical systems Xa (the source solution) having the goal "G': TO_SENSE_RATE_of v: voltage CHANGE". Goal G' is considered "similar" to G* since both share a common generalization G. Given the exemplar P1, additional knowledge is derived about the source solution by following links top down from teleology to function and selecting its functional representation. The functional fragment specifies that the goal "G': TO_SENSE_RATE_of v: voltage CHANGE" is realised in this device by a physical process of the type "Reservoir charging" in the electrical domain whose cofunction is constituted by three functional roles - namely, a voltage generator, a conduit of current and a reservoir of charge - related by dependency links. The desired output (i.e., dv/dt) is measured across the conduit which is equivalent to a generator when it is in the functional state "crossed". Given the functional fragment for the class of systems Xa, this model is mapped over to the target domain (i.e., in the hydraulic domain) to represent the class of systems Xb. Thus, voltage, current and charge are mapped into pressure, volume flow rate and volume (of gas) respectively. As a consequence, the source cofunction in the electrical domain is mapped to a similar cofunction, in the hydraulic domain, constituted by a generator of pressure, a conduit of flow and a reservoir of volume (of gas) related by dependency links. Finally, given mapped target cofunction, the last step consists in searching the library of exemplars to find components that are capable of playing the roles considered in the cofunction. Since, the component which plays the role of the generator of pressure is known from the specification only the other two roles - namely, the conduit of flow and the reservoir of gas - must be instantiated. The retrieval process provides a fluid resistance (e.g. an orifice) and a fluid capacitance (e.g., a fluid accumulator) as candidate components. The design solution is thus constituted by composing the generator of pressure specified in input with an orifice and a fluid accumulator. The desired output (i.e. dp/dt) must be measured across the orifice. The plausibility of this conclusion rests on the assumption that the type of goal determines (or is relevant to) the functional organisation of a device independently of physical domains. Of course, this may be incorrect but often this heuristic provides a good starting point for the conceptual design stage.

An accurate analysis of the foregoing line of reasoning reveals a complex pattern of inferences including similarization, retrieval, conceptual abstraction and composition. Some of them (e.g. similarization, composition) are performed explicitly as a sequence of one or more individual steps. Others (e.g. conceptual concretion) are, in fact, compiled into a case or a prototype and thus are executed implicitly by recalling the case or the prototype from background knowledge (i.e. by case or prototype based reasoning). The strengths and weakness of both approaches are well known. Explicit use of model transmutations rapidly becomes intractable as the complexity and the magnitude of the problem increase. However, since this approach is based on "first" and "second" principles it is especially suited for solving novel problems (i.e. for innovative and creative design). On the other hand, case-based reasoning or prototype-based reasoning since they reason from previous experience may improve productivity and guarantee standardised solutions when applied to solve routine problems. However, their effectiveness is strongly limited if new problems or problems requiring some form of creative processing are addressed. In real design problems a tradeoff between efficiency and effectiveness is expected: some parts of the operational model are generated implicitly by recalling from memory past cases and by adapting these cases using transformations to the present situation while other parts of the operational model are generated by constructing and composing fragments using model transmutations explicitly.

8. Relationship among input information, prior design knowledge and model transmutations

Two critical factors affect the design process: the "complexity" of the initial design state and the relation existing between design states and background knowledge.

The "complexity" of the initial design state can be characterized along two dimensions: amount of unspecified structure and gap in abstraction level between input specification and desired solution. Because a design problem often provides part of the solution it is useful to identify what of the desired artifact's structure is left to be determined. A design problem, for example, may specify the topology of the desired artifact letting the designer to determine specific values for system parameters or may specify the components constituting the desired artifact but not their connections, or, finally, may require the designer to synthesise new components. As far as the second dimension is concerned, it should be noted that in the simplest case design specification is formulated at the same level of abstraction (e.g., structural or behavioral) than the desired solution. In the worst case specification is expressed at an abstraction level (e.g. teleological) far separated from the level of solution (i.e. structural) thus requiring several conceptual transformations between intermediate levels to build the final solution. The "complexity" of the starting design state DS(0) affects the choice of the initial transmutation. This step, also known as incipient model creation, is very important since a model that is created first has a large influence on the result.

The second factor concerns the relation between design states and background knowledge. At each step of the design process the current design state activates portions of the background knowledge which are relevant to the current specification or operational model. Notice that the inclusion of contextual information in design fragments and the use of concept hierarchies in organising background knowledge facilitate this tasks. For example, by explicitly representing the ontology of a model it is possible to recall the concept hierarchies that are conceptually relevant for that model i.e. those that match the entities and relations specified in the model. The selection of what action to do next depends on the relationships between relevant background knowledge and the information that is provided in the current design state. Several relationships can be envisaged:

1) The current state is already known to the designer. This case occurs, for example, when design goals match exactly some part of BK (e.g. the teleological fragment of a prototype). In such a situation, the stored prototype is recalled and included into the current operational model.

2) The current state is implied (or implies) part of BK. This case occurs, for example, when a part of BK accounts for the input specification. Consider, for example, the problem of designing a system which can achieve goals G1* and G2*. Suppose further that the BK does not contain a prototype that fulfils both goals but contains two

prototypes P1 and P2 each one partially matching the given functional specification (i.e. the desired goals G1* and G2* appear as subgoals in the teleological fragments associated to prototypes P1 and P2 respectively). By exploiting the links existing between design fragments of P1 and P2 it is possible to isolate the parts of the prototypes that are responsible for the achievement of the specified goals, include these fragments in the current operational model and then try to compose these fragments together to obtain a possible model of the desired artifact. Notice that composition can be performed at the behavioral or functional levels.

3) The current state evokes an analogy to a part of BK. This case represents the situation when the design goal does not match any background knowledge nor it is implied by it. Similarization can thus be used to generate a similar goal that is used to retrieve the operational model of an analogous system. This model is in turn transformed, again by similarization, in the desired physical domain as illustrated in the previous section.

4) The current state represents completely new information. A solution is not known or cannot be discovered by search but it must be generated from scratch (explorative design) using basic transmutations. The designer may try to generalize (or specialise) the input specification, relax some goals or constraints, negate desired goals, etc. so that the new specification can account for information stored previously and the control can be passed to one of the above cases.

Examples of cases 2) and 3) are discussed in detail in [21].

9. The SECS system

In order to experiment with the above ideas we developed SECS[1] a reasoning system aimed at supporting the electronic engineer in the phase of conceptual design of small electronic circuits. SECS has been implemented on Apple Macintosh machines using LPA Prolog. The system allows the designer to use either past experiences in the domain of application or basic transmutations to adapt or generate design solutions. Past experiences are recorded by means of a library of design prototypes which are organized into generalization (type-of) and meronimic (part-of) hierarchies. This type of organization is more or less standard but we identified specific levels of generalisation - namely, teleological class, design concept, variant, plan, schema and case - according to the extension of the reference set described by a prototype. Prototypes are indexed by the goals they achieve and the operational conditions under which it is appropriate to use them. Notice that this choice corresponds to view the teleological description of a prototype as the specification of the design problem it may resolve while the other descriptions included in the prototype (i.e. the functional, behavioral and structural fragments) represent a possible solution of the problem at different epistemological levels. The operation of SECS is based on the classical Retrieve/Select/Adapt method [20]. In the current implementation of the system there is not an explicit decision phase. Switching from a strategy to another (e.g. from prototype-based reasoning to the explicit use of model transmutations) is failure-driven. It occurs when: i) there are no prototypes satisfying one or more of the specified goals, ii) there are two or more competing prototypes satisfying only a subset of specified goals and these subsets have some element in common. In case i) the system switches to cross contextual analogical reasoning using similarization. In case ii) (corresponding to case 2 in the previous section) the system switches to a compositional based strategy using partial prototypes represented at the functional level as elementary fragments. Composition-based design (at the functional or behavioral levels) is performed

---

[1] SECS is an acronym for Small Electronic Circuit Synthesizer.

also when initial design specification is formulated in functional or behavioral terms. More detail about the SECS system can be found in [1], [21] where several examples of its functioning are illustrated. It should be noted that the control mechanism of the current implementation of the SECS system is not truly adaptive in the sense described in section 5. Current research efforts are aimed at implementing the complete decision step (i.e. analysis and evaluation, critique, and selection) by adapting the method proposed in [22] for task-adaptive model selection to the choice of model transmutations.

10. Conclusions and related work

In this paper we propose to view design as a bilevel reflective process. This process is constituted, at the base level, by a sequence of modeling steps which are dictated, at the metalevel, by a decision activity which continuously monitors the progress of the design process at the base level and determines at run time what to do next and when to stop. The modeling activity has been analysed in terms of patterns of inference called model transmutations. Four categories of transmutations have been discussed with reference to the Multimodeling approach for representing physical systems. These are: transformations, combinations and retrievals. The major goal of the paper is to provide a conceptual framework for analysing design systems and for addressing questions concerning their competence such as what types of inference patterns underlie different design strategies e.g. top-down, compositional and analogical design; what kind of design solutions a design system is able to generate from what kind of input specification and prior design knowledge; what is the logical relationship between specification and background knowledge and so forth.

The analysis of model transmutations builds on earlier research in compositional model-based design [3], innovative design [17], [24], prototype-based reasoning [7], [19] and cross contextual analogical reasoning [9], [16]. The work presented in the paper also integrates several results obtained in the area of qualitative physics and automated modeling [6], [10], [15], and multistrategy task adaptive learning [12]. The major contributions of our research derive from the use of the Multimodeling approach for representing both design experience (i.e. cases and prototypes) and design solutions (i.e. design fragments and operational models). The approach provides a set of specific ontologies for representing all epistemological types (i.e., structural, behavioural, functional and teleological) and a systematic method for building different models of the same artifact (e.g. models having different epistemological type, different coverage, resolution or detail) and for relating all these models by links.

References

[1] Beltrame A. and Toppano E. (1995). Prototype-based conceptual design: the SECS system. In *Applications of AI in Engineering X*, R.A. Adey, G.Rzevski, and C.Tasso (Eds.), Comp. Mechanics Pub., Boston, pp. 502-512.

[2] Borst P., Akkermans H., Pos A., and Top J. (1995). The PhysSys Ontology for Physical Systems. *Proc. QR Workshop*, Amsterman, the Netherlands, pp. 11-21.

[3] Bose P. and Rajamoney S.A. (1993). Compositional model-based design. *Proc. IJCAI-93*, Chambery, France, pp. 1445-1450.

[4] Chandrasekaran B. (1990). Design problem solving: a task analysis, *AI Magazine*, Winter 1990, pp. 59-71.

[5] Chittaro L., Guida G., Tasso C. and Toppano E. (1993). Functional and teleological knowledge in the Multimodeling approach for reasoning about physical systems: a

case study in diagnosis. In *IEEE Transactions on Systems, Man, and Cybernetics*, vol. 23, No.6, pp. 1718-1751.

[6] Falkenhainer B., and Forbus K.D. (1991). Compositional Modeling: Finding the Right Model for the Job, *Artificial Intelligence* 51, pp. 95-143.

[7] Gero J.S. (1990). Design prototypes: a knowledge representation schema for design. In *AI Magazine* 11 (4), Winter 1990, pp. 26-36.

[8] Goel A., and Chandrasekaran B. (1989). Functional Representation of Designs and Redesign Problem Solving. *Proc. IJCAI-89*, Detroit, MI, USA, pp. 1388-1394.

[9] Greiner R. (1988). Learning by understanding analogies. *Artificial Intelligence*, vol. 35, pp. 81-125.

[10] Iwasaki Y., and Levy A.Y. (1994). Automated model selection for simulation. In *Proc. AAAI-94*, Seattle, WA, USA, pp. 1183-1190.

[11] Kannapan S.M. (1993). Metrics for functional evaluation of engineered devices. In *Proc. of the Reasoning about Function Workshop*, July 1993, Washington, D.C., pp. 53-59.

[12] Michalski R.S. (1991). Inferential learning theory as a basis for multitrategy task-adaptive learning. In *Proc. First International Workshop on Multistrategy Learning*, November 1991, Harpers Ferry, West Virginia, pp. 3-18.

[13] Michalski R.S., and Hieb, M.R. (1993). Knowledge representation for Multistrategy Task-Adaptive Learning: Dynamic Interlaced Hierarchies. In *Proc. Second International Workshop on Multistrategy Learning*, Harpers Ferry, West Virginia, pp. 3-17.

[14] Murthy S. (1988). Qualitative reasoning at multiple resolutions. In *Proc. AAAI-88*, Saint Paul, MN, USA, pp. 296-300.

[15] Nayak P.P. (1994). Causal approximations. *Artificial Intelligence*, vol. 70, pp. 277-334.

[16] Navinchandra D., Sriram D., and Kedar Cabelli S.T. (1987). Analogy-based engineering problem solving: an overview. In *AI in Engineering: Tools and Techniques*, Sriram &Adey (Eds.), Computational Mechanics Pub., pp. 273-285.

[17] Neville D., and Weld D.S. (1993). Innovative design as systematic search. In *Proc. AAAI-93*, Washington, D.C., USA, pp. 737-742.

[18] Pahl G., and Beitz W. (1984).*Engineering Design*, The Design Council, London,.

[19] Rajamoney S.A. and Lee H. (1991). Prototype-Based reasoning. In *Proc. AAAI-91*, Anaheim, CA, USA, pp. 34-39.

[20] Riesbeck, C.K. and Schank R.C. (1989). *Inside Case-Based Reasoning*. Lawrence Erlbaum Associates Pulishers, Hillsdale, New Jersey.

[21] Toppano E. (1996). Multistrategy modeling: a case study in design. In *Proc. 6th European-Japanese Seminar on Information Modeling and Knowledge Bases*, Hornbaek, Denmark, pp. 35-48.

[22] Toppano E. (1996). Rational Model Selection in Large Engineering Knowledge Bases. *Applied Artificial Intelligence Journal* 10, No. 3, pp. 191-224.

[23] Umeda Y., Takeda H., Tomiyama T., and Yoshikawa H. (1990). Function, Behaviour, and Structure. In *Applications of Artificial Intelligence in Engineering V*, vol.1, J.S.Gero Ed. Springer-Verlag, pp. 177-193.

[24] Williams B.C. (1990). Interaction-based invention: designing novel devices from first principles. In *Expert Systems in Engineering: Principles and Applications. Proc. Int. Workshop*, Vienna, Austria, pp. 119-134.

[25] Wills L.M., and Kolodner J.L. (1994). Towards more creative case-based design systems. In *Proc. AAAI-94*, Seattle, Washington, pp. 50-55.

HANDLING REGULATORY INFORMATION
USING INFORMATION TECHNOLOGY

P.W.H. Chung
Loughborough University of Technology, Loughborough, UK

D. Stone
Building Directorate, Scottish Office, Edinburgh, UK

Abstract

Many aspects of some industries, like the construction and process industries, are governed by regulatory information. This information has a direct and significant effect on the safety, economics and quality of the industries' operations and end-products. Regulatory information has long been recognised as a potentially rich area for computer applications. This paper provides an overview of the work that has been done to provide computer systems that will help both users and authors of regulations. The emphasis is on using advanced information technology that would help in accessing, interpreting and applying the information.

1. Introduction

Many aspects of some industries, like the construction and process industries, are governed by regulatory information. This information has a direct and significant effect on the safety, economics and quality of the industries' operations and end-products. The information is contained in a range of document types extending from the mandatory to the advisory and includes national and local regulations, codes of practice and product standards. For the purpose of this review, the terms *regulations, codes of practice*, and *standards* are used interchangeably.

Standards are often voluminous and can be complex, consisting of contributions from a number of authors. The information is costly to produce in terms of both time and effort. It requires the involvement of leading experts in the subject domain and, usually, protracted consultation and review procedures which lead to a cycle of revision and re-drafting.

The volume of information involved compels individual users to invest considerable time and effort in locating relevant parts. Once located there are often further difficulties associated with understanding and correctly applying the requirements. Idiosyncrasies of style and inadvertent inconsistencies or errors in the texts can, in the case of mandatory requirements, lead to conflicting interpretations by users and enforcing authorities.

Legislation and similar regulatory information has long been recognised as a potentially rich area for computer applications. Conventional legislation, however, has been found to contain many structural and conceptual problems which defy easy analysis and representation (Gardner, A. 1987; Susskind, R. 1987). However, standards in technical domains do not share these problems to the same extent because they focus on the physical rather than the abstract. This paper reviews work carried out related to technical standards, in particular building standards because it is in this domain that most of the research has been done to date.

Section two of the paper considers the provision of on-line access to users. Different aspects of domain modelling, i.e. providing a framework so that information can be represented in a systematic and formal way, is considered in section three. Section four looks at different approaches to implementing knowledge-based standards processors. The idea of automated text formalisation is briefly described in section five. Finally, recent work on providing support tools for authors of standards is discussed in section six.

2. Retrieval Systems

Because of the volume and complexity of standards, a user, who may be a designer or a local authority official whose job is to administer the standards, has to invest considerable effort in order to understand the inter-relationship of the different parts. To facilitate user access, regulatory information is now being put onto computer databases (Vanier, 1991; Bourdeau, 1991). Irrespective of what hardware is used to store the information, a hypertext like interface for browsing is preferred. It allows the user to move from one piece of information to another piece of information relatively easily. This facility is desirable because it is common that one part of a document references other clauses, tables, figures, graphs and equations, as well as other documents.

The main access problem lies in how to index an entire document or a set of documents so that the user can search for the relevant information without being overwhelmed by the volume of material available. There are basically two approaches. One is to use keyword descriptors and the other is a structured approach based on a classification of the concepts in the domain.

2.1 Keyword Descriptors

One of the largest scale projects in providing on-line access to regulatory information was carried out by Bourdeau and colleagues (Bourdeau, 1991). REEF is a 15000 page encyclopedia consisting of a collection of about 1000 documents commonly used by French building professionals such as architects, engineers or contractors. In order to facilitate information retrieval REEF has been put on a CD-ROM to make the information accessible from a standard microcomputer. Texts make up 75% of the database and the rest consist of 3500 tables, 9000 drawings and a large number of formulae. A huge amount of effort was spent digitizing all the documents. The first release of the CD version of REEF (CD-REEF) was made available at the end of 1991.

Information contained in CD-REEF is indexed at two different levels: the document level and the information unit level. At the former level, the content of each document is characterized by two types of descriptors: general and technical. A general descriptor has four attributes: works, functional requirements, administrative constraints and purposes. A technical descriptor has two attributes: products and materials. Figures 1 and 2 are examples taken from Bourdeau (1991). The descriptors are used to index a document about roofing using large sheets of stainless steel.

	Works	Functional requirements	Administrative constraints	Purposes
Descriptor 1	roofing	mechanical resistance and stability	builder obligations	all buildings
Descriptor 2	roofing	mechanical resistance and stability	owner obligations	all buildings
Descriptor 3	roofing	water tightness	builder obligations	all buildings
Descriptor 4	roofing	water tightness	owner obligations	all buildings
Descriptor 5	roof window	water tightness		all buildings

Figure 1 An Example General Descriptor

Products	Materials
metallic sheets	stainless steel

Figure 2 An Example Technical Descriptor

At the second level, each document is divided into information units. Each unit is indexed by a set of keywords that describe its content. The keywords are first of all extracted automatically from the text. They are then checked by human experts and inappropriate words are deleted. The dictionary of keywords has about 5000 entries. Future work includes building a thesaurus so that a search can be extended to include synonyms and associated terms.

Accessing information from CD-REEF involves two steps, each corresponding to a different level of indexing.

Step 1: select document. This can be done by selecting directly from the catalogue of documents or by specifying general and technical descriptors to initiate a search. After the appropriate documents have be selected, the user can copy them to a file specific to the case in progress.

Step 2: select information unit. Again there are two ways of doing this. The first is by looking up the table of contents of the selected documents and identifying the sections that are of interest. The second is by specifying keywords. The user can choose the keywords from a list, which is generated by the system by merging the indexes of all the information units in the selected documents.

It is not clear how easy it is for a user to locate the relevant parts of the regulation using CD-REEF. A potential problem in step one is for the user to provide the appropriate general and technical descriptors, particularly if there is no guidance given by the system. A problem related to the use of keywords in step 2 is that the system has no idea how different keywords are related to one another. For example, if the user is interested in looking at regulations related to all types of heating system then he or she may have to provide an exhaustive list of keywords. There is also the risk of missing some of the keywords out.

2.2 Classification Scheme

Some researchers (Harris and Wright, 1980) see that the way to understand the inter-relationships of the concepts contained in codes is by creating a formal structure of those concepts. Vanier (1991) describes an exploratory project that attempts to develop a classification system for the National Building Code of Canada (NBC). The NBC Classification System has three components: a classification scheme, an article database and a retrieval interface.

The classification scheme is essentially a network of nodes and links. A node represents a concept that is found in the NBC. A link represents the relationship between the two nodes that it joins together. Figure 3 shows a partial network related to the concept "service water heater". The classification scheme is being developed by experts using HyperCard. Based on counting the number of discrete words, the NBC has roughly 2000 to 3000 concepts. It is estimated that it will take approximately one year to construct the complete classification scheme for the NBC.

In the Article Database, each information unit is carefully indexed by the relevant concepts. Access to the database is via the Retrieval Interface. The user specifies the concepts which describe the project at hand and the interface program extracts the relevant information units from the article database. Because the system has an explicit model of the relationships among the concepts it will return not only those units that exactly match the specification but also their sub-categories. For example, the user could quite easily ask for information that relates to gas fires specifically or heating systems in general. In the latter case, information about gas fires would be included in the information returned by the system.

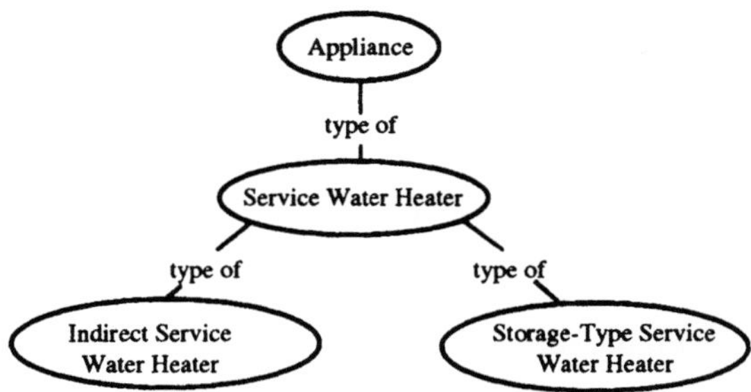

Figure 3 Service Water Heater Hierarchy

3. Domain Modelling

In the previous discussion of retrieval systems, we have briefly touched on the idea of classification but have not addressed properly the more general concept of domain modelling. Fenves et al (1987) argued that the systematic representation and manipulation of regulatory information by computers cannot be approached in an ad-hoc manner. Standards are, in their entirety, a substantial body of information and cover a wide range of issues. There is a common consensus that clearly specified domain models will greatly facilitate software development and integration.

However, the term model is very vague and all encompassing. In the most general sense, models are collections of concepts linked by some kind of relationships. Within a domain, there are different kinds of models and they are used for different purposes. Different kinds of models are not to be confused with different modelling tools. There is a variety of tools available but they will not be discussed here. Kahkonen et al (1991) provides a comparison of three modelling tools EXPRESS-G, IDEF1X and NIAM. The concern here are the different types of models and the ways that they are used. Five different uses of models have been identified. They are type hierarchy, part hierarchy, data dependency, requirements and semantics.

3.1 Type hierarchy

Type hierarchy is used to show, as the name suggests, the "type of" relation. A type hierarchy begins with a general concept as the root node. Each branch from the node represents a specialisation of the concept. A concept can be either physical or abstract. Figure 3 in section 2.2 is a type hierarchy of physical objects. Figure 4 is a simple hierarchy of abstract concepts.

Associated with each concept is a set of attributes. For example, the concept "wall" may have the attributes: thickness, height, width and material. Different concepts may have the same attribute names associated with them. For example, the concept "building" may also have the attribute "height". A real object, for example a particular building or a particular heater, is an "instance" of a concept in a type hierarchy. This is also known as the object-oriented approach to organising information. It provides a convenient way of grouping information together. As mentioned earlier, a classification scheme that is based on type hierarchies allows the user to request information either at a general or a specific level.

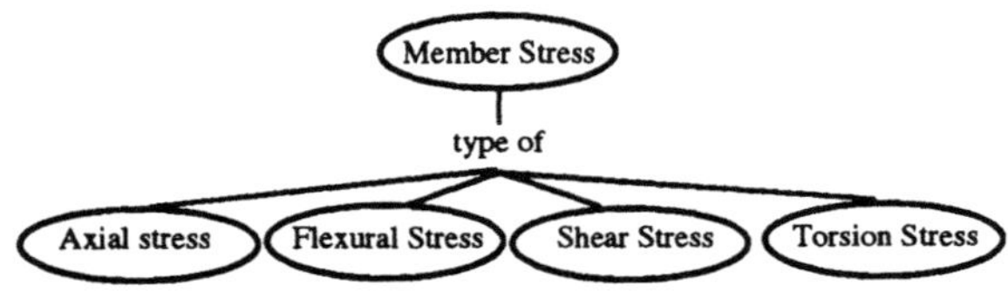

Figure 4 Member Stress Hierarchy

3.2 Part hierarchy

Another important relationship that needs to be captured is the "part of" relation. A part hierarchy is to show the constituent parts that make up a concept. For example, a house may have the following parts: window, door, bedroom, kitchen, toilet. Individual requirements that apply to houses refer to one or more parts of the house. A part hierarchy shows *what* needs to be considered.

3.3 Data dependency

Information networks are constructed to show how data in the domain are dependent on one another (Fenves et al, 1987). An ingredience network shows, for any given conclusion of a provisional clause, all the items of information and their precedence which contribute to that conclusion. For example, figure 5 shows that the determination of the aggregate width required for escape stairs requires the determination of both the calculated width and the minimum width required for all stairs. The latter requires the calculation of the appropriate capacity of the storeys served and the determination of the minimum number of stairways required. Dependence networks are the complement of ingredient networks and show, for any given condition, all the conclusions to which it contributes. For example, the height of a storey above ground level is a common ingredient that affects many conclusions.

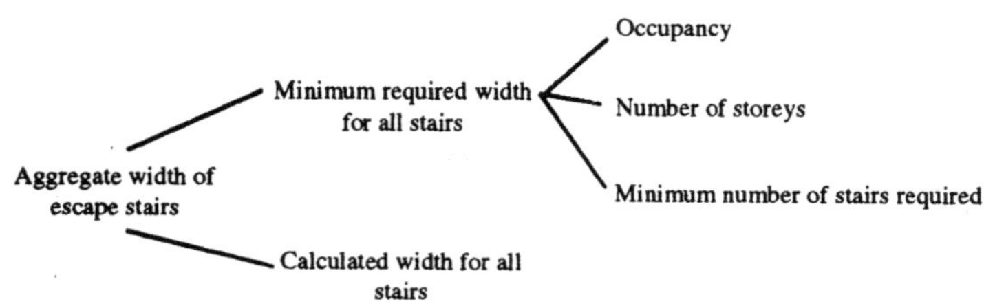

Figure 5 An Example Ingredience Network

There are a number of advantages to this kind of analysis of information which may benefit both authors and users. An ingredience network reveals the procedural implications of the information and may suggest possible orderings for material. A dependence network can show all the requirement conclusions which may be critically affected by a design variable.

3.4 Requirements

Closely related to information network is the modelling of the individual requirements in Standards. The aim here is not only to show data dependency but also to show *how* data are actually dependent on one another. The earliest work on representing the logical structure of provisions was by Fenves (1966). It was based on the ideas of decision tables and decision trees. Since then a considerable amount of work has been done to demonstrate that these ideas are practicable (Fenves et al, 1987). Other researchers have tried to represent requirements using logical rules. However, their work are more tied to implementation than modelling. Therefore, discussion of rules will be deferred until the section on expert systems.

3.4.1 Decision tables

A decision table has three parts: a set of conditions, a set of actions and a set of rules. Each rule shows the actions to be taken for a particular combination of conditions. Figure 6 shows the decision table representation of the following building regulation statement taken from Stone and Wilcox (1988):

> If   the occupancy sub-group of the building is A1
> and either
>   1) the element of structure is a separating wall or;
>   2) the height of the building is greater than 28 metres or;
>   3) the element of structure is a compartment wall and the height of the
>   building is greater than 15 meters
> then the minimum fire resistance of the element of structure is 60 minutes;
> in any other case the fire resistance is 30 minutes.

		Rule 1	Rule 2	Rule 3	Rule 4	Rule 5
Conditions	The occupancy sub-group of the building is A1	True	True	True	True	True
	The element of structure is a separating wall	True				False
	The height of the building is greater than 28 metres		True			False
	The element of structure is a compartment wall			True	True	False
	The height of the building is greater than 15 metres			True	False	
Actions	The minimum fire resistance of the element of structure is 60 minutes	Yes	Yes	Yes		
	The minimum fire resistance of the element of structure is 30 minutes				Yes	Yes

Figure 6 An Example Decision Table

A careful analysis of the statement will show that there is an ambiguity about what "in any other case" means. It could mean two things, either if the occupancy sub-group of the building is not A1, or the occupancy sub-group of the building is A1 but not 1), 2) or 3). Here it is taken to mean the latter.

Each rule in the table suggests a particular conclusion for a particular combination of conditions. For example rule 1 is interpreted as: if the occupancy sub-group of the building is A1 and the element of structure is a separating wall then the minimum fire resistance of the element of structure is 60 minutes.

Using this approach, a document will be represented by a set of inter-related tables. The tables are inter-related because the values of variables contained in the condition part of one table may be derived from the conclusions of other tables. Information networks as discussed in the last section can be generated from the decision tables automatically.

3.4.2 Decision trees

Another representation formalism of requirements is the decision tree. For example, figure 7 shows a partial decision tree of the statement given in the previous section. A decision tree does not need to be constructed from the original statement. It can be easily translated from a corresponding decision table. It is worth noting that for a given decision table there can be many valid decision trees. Depending on the order in which the conditions are tested, some decision trees are smaller than others.

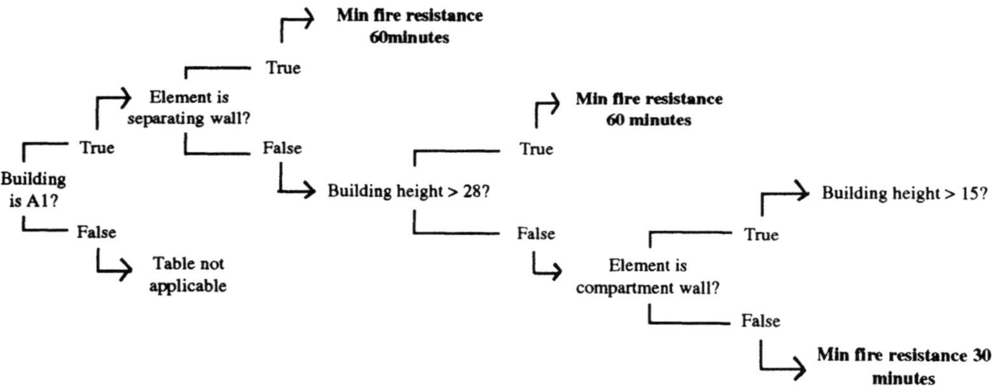

Figure 7 A Partial Decision Tree

3.5 Semantics

Some researchers take modelling to the extreme. Every relationship identified in a Standards has to be made explicit so that every statement can be representing formally as a graph (Cornick et al, 1991). This is to ensure that there is no ambiguity in the standards. Figure 8 is a partial model for the sentence "A fire-separation means a construction assembly that acts as a barrier against the spread of fire". However, it is not entirely obvious that the model is clearer than the original sentence. More complicated examples are given in Cornick et al (1991). This suggests that there is a limit to what one can model for practical purposes. Beyond a reasonable limit the modelling activity will bring very little benefit, if any at all.

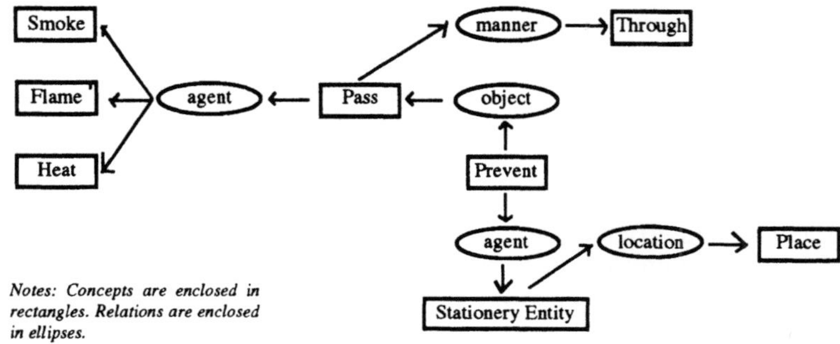

Notes: Concepts are enclosed in rectangles. Relations are enclosed in ellipses.

Figure 8 Partial Model for Fire Separation

3.6 Validation of logical properties

Much regulatory information is prescriptive in character. It sets out, for a given set of conditions, the required actions. A number of researchers, notably Fenves et al (1987), have pointed out that, given this prescriptive intent, it is critical that the information exhibits a number of important logical properties in order to avoid difficulties of interpretation and application in practice. They argue that the information should be complete, unique, acyclical and non-redundant. The kinds of representations described in previous sections which are suitable for prescriptive information, that is, decision logic tables, decision trees and production rules, lend themselves to formal analysis and validation to determine if these required logical properties obtain. In addition, these representations are susceptible to topological analysis to construct the data dependency networks described in section 3.3 above.

4. Expert Systems

Besides making standards more accessible by providing on-line retrieval facilities, some projects have attempted to encode the standards themselves as executable programs. This kind of program takes a set of design parameters as input and help to speed up the design process by automating a number of tasks: determine requirements, perform calculations, carry out compliance checks. For example, Fire Expert (Tuominen, 1991) is an expert system for Finnish fire safety regulations; ENVSTD (Crawley and Boulin, 1991), which stands for ENVelope STanDard, is a program for complying with building envelope requirements based on America's "Energy Efficient Design of New Buildings Except Low-Rise Residential Buildings" Standard 90.1.

There are different ways of implementing standards processors. One way is to write the programs using conventional procedural languages. For example EVNSTD is implemented in C. However, converting a Standards document into an executable program requires a lot of programming effort. Furthermore, maintenance of the program to reflect changes in the document is also a problem. Therefore, some researchers have experimented with high level programming tools that might facilitate the development of such programs. This section reviews a number of different approaches that have been proposed and tested to a limited extent.

4.1 Logic programming

The objective of logic programming is to provide a means of defining computer applications in terms of a machine intelligible form of symbolic logic. Symbolic logic is based on propositional logic which is concerned with expressing propositions and the relationships between propositions and determining how one proposition can be validly inferred from another; a proposition in this sense is simply a statement which may be either true or false.

The most practical and efficient implementation of the notion of logic programming at the moment is the programming language Prolog (Clocksin and Mellish, 1984). A Prolog program consists of a set of Horn clauses, which are rules of the form

Conclusion if *Condition 1* and ..... *Condition N*.

This means that *Conclusion* is true if it can be shown that *Condition 1* to *Condition N* are true.

It is quite obvious that there is a close mapping between the general form of a Horn clause and a requirement. The regulation statement considered previously can be written as a complete running Prolog program as shown in figure 9. Each clause in the Prolog program corresponds to a column in the corresponding decision table. Each symbol that begins with a capital letter is a variable. When Prolog is presented with the goal of the form *min-fire-resistance(element,X)*, it will deduce the *min-fire-resistance* rating for *element* and bind the value to variable X.

Although Prolog provides a very convenient representation and a powerful computation mechanism, it is not without problems in practice. Notice that some of the conditions are duplicated a number of times in the rules. This may present two problems. One is consistent updating. If one condition in the Standards is amended it will need to be changed in several places in the program. The other problem is slow execution because the same condition has to be evaluated several times. One could rewrite the program in other ways by taking advantage of the control facilities provided in Prolog. However, this would obscure the meaning of the program. Therefore, in practice there is a trade off between clarity and speed.

Another aspect of the sample program that requires special attention is the use of the *not* predicate in the fifth clause. The goal *not(Q)* in Prolog is interpreted as *all ways of showing Q fail*. This interpretation is called negation as failure (Clark, 1978). It is justified whenever we can make a *closed world assumption*: anything which is not known is assumed to be false. However, not making a clear distinction between *not known* and *false* will sometimes lead to wrong inferences. For example, if the type of wall is not known the sample program will falsely infer that the min-fire-resistance of the element is 30 minutes. This problem may be serious because regulations are often drafted by general rules followed separately by a list of exceptions. Sergot et al (1986) discusses the treatment of negated conditions in more detail using examples from the British Nationality Act.

Prolog's in-built control strategy entails backward reasoning, that is, it reasons from conditions to a conclusion. The normal query form is to ask the system to prove or find a value for a rule conclusion. In some situations, however, it is useful to be able to specify a preferred conclusion and ask the system to determine what are the necessary condition values. Stone and Wilcox (1986) discusses some of the problems of representing regulations using logic in more detail. Despite the criticisms, Prolog, compared with other programming languages, does provide a convenient way of implementing a Standards processor. However, a significant amount of programming skill and effort is still required.

```
min-fire-resistance(E,60) if              % The minimum fire resistance of E is 60 minutes
    element-of-structure(E,B), building(B),    % E is an element of structure of B and B is a building
    occupancy-sub-group(B,a1),                 % The occupancy sub-group of B is a1
    separating-wall(E).                        % E is a separating wall

min-fire-resistance(E,60) if
    element-of-structure(E,B), building(B),
    occupancy-sub-group(B,a1),
    height(B,H), H > 28.                       % The height of B is H and H is greater than 28 meters

min-fire-resistance(E,60) if
    element-of-structure(E,B), building(B),
    occupancy-sub-group(B,a1),
    compartment-wall(E),                       % E is a compartment wall
    height(B,H), H > 15.                       % The height of B is H and H is greater than 15 meters

min-fire-resistance(E,30) if                  % The minimum fire resistance of E is 30 minutes
    element-of-structure(E,B), building(B),
    occupancy-sub-group(B,a1),
    compartment-wall(E),                       % E is a compartment wall
    height(B,H), H =< 15.                      % The height of B is H and H is less than or equal to 15 meters

min-fire-resistance(E,30) if
    element-of-structure(E,B), building(B),
    occupancy-sub-group(B,a1),
    not separating-wall(E),                    % E is not a separating-wall
    not compartment-wall(E).                   % E is not a compartment-wall
    height(B,H), not H > 28,                   % The height of B is H and H is not greater than 28 meters
```

Figure 9 A sample Prolog program

4.2 Production rules

Production system (Brownston et al, 1985) provides another way of implementing standards processors. Syntactically, a production rule is very similar to a clause in Prolog:

If Condition 1 and .... and Condition N then Conclusion

Here the conditions are specified before the conclusion. However, the main difference lies in the way the rules are tested for execution. A production system works in a data-driven manner rather than in a goal-directed manner as in Prolog. Whenever all the conditions of a rule are satisfied, i.e. there are data in the system that match the all conditions of the rule, the actions specified in the conclusion part is executed automatically. Therefore, to control the execution of the program, control rules need to be added and the condition part of the domain rules also need to be extended to include control information. As discussed in the previous section, when control information has to be included in a rule-based system, the meaning of the rules become less clear. Updating and modifying the system to reflect changes in the standards is much more difficult. Therefore, production systems also share many of the problems that Prolog has.

4.3 Generic standards processors

To overcome the problem of mixing domain and control information as discussed in the previous section, some researchers have proposed the development of a generic standards processor for the building domain (Rosenman and Gero, 1985). The aim is to provide a tool that allows the user to input the domain knowledge in a purely declarative way. The written text should map to the formal representation easily and the user should not be concerned about how the program is going to run. Three different representations have been tried: decision tables, rules and facts.

Decision tables, as discussed previously, are very convenient for representing the logical structure of requirements. Decision tables can be translated into decision trees which can then be executed. However, tables only provide a high level framework for structuring information. They do not specify how conditions and conclusions should be expressed. It is not clear from the literature (Garrett and Fenves, 1986; Fenves et al, 1987) what language has been developed for the user.

Rosenman and Gero (1985) consider rules as an appropriate representation for Standards. Recognising the disadvantages of using a general purpose language like Prolog or an expert system shell, they have developed their own shells. However, it is difficult to work out from the literature exactly what facilities their system provides in addition to a general purpose rule-based shell.

Other researchers (Rasdorf and Wang, 1988; Topping and Kumar, 1989) have suggested representing Standards requirements as a collection of facts. The facts are then interpreted by a generic standards processor as and when they are needed. The general idea behind this approach is to divide a requirement into one or both of the following two types of criteria: applicability criteria and performance criteria. They are analogous to the conditions and actions of a rule respectively. The general forms for the applicability and performance criteria are:

applicability(Clause Number, Expression)
performance(Clause Number, Data Item, Expression, Operator)

The outcome of converting the example building regulation statement into facts would be something like figure 10.

```
applicability(3.2, occupancy-sub-group(B,a1))

applicability(3.2, height(B) > 15)

applicability(3.2, height(B) > 28)

applicability(3.2, separating-wall(E))

applicability(3.2, compartment-wall(E))

performance(3.2, min-fire-resistance, 30, =)

performance(3.2, min-fire-resistance, 60, =)
```

Figure 10 Example Applicability and Performance Criteria

An analysis of the facts in figure 10 reveals two fundamental problems with this approach. The first problem is the treatment of disjunctions. The use of a single clause number is insufficient to distinguish how the different applicability criteria relate to the different performance criteria. This may be overcome by creating artificial clause numbers. The second, and more serious, problem is the treatment of variables. Because facts are represented individually there is no information linking that the element E is part of the building B. It seems that there is no advantage of representing the applicability and performance criteria as separate facts.

Some of the work carried out so far is encouraging. It has been demonstrated that expert systems have a role in standards processing. However, there is no generic processor that has made it beyond the research laboratory. There is still much research to be done before such a tool will be made available. The rule-based approach seems intuitive and natural. However, decision tables provides a much more concise and compact structure. From our analysis we conclude that development of decision table based software using an AI language, like Prolog, with object-oriented extension is a promising way forward.

5. Automated Formalisation of Texts

Stone and Wilcox (1987) have sought to devise methods which directly formalise and map regulatory information into a machine readable and executable form. They suggested that there are two advantages to automated formalisation of texts. Firstly, it obviates the need for intermediary specialists or knowledge engineers to translate the information into machine format. Secondly, it provides the possibility for authors of codes and standards to communicate directly with system functions in familiar text form without the need to understand system notions such as predicates, arguments, variables and so forth.

Automated text formalisation depends essentially upon some form of parsing. It seems possible that regulatory information, because of its formal function, can be expressed in a form of structured English which uses a limited lexicon of words and phrases. Stone and Wilcox (1987) describe a parser which anticipates a restricted grammatical structure and a domain specific lexicon. Stone and Wilcox (1988) describe a simplified parser which makes use of a limited set of logical connectives and operators. Text strings between connectives are treated as logical constants except that certain keyword operators, typically those representing mathematical functions, are replaced with the equivalent system function and a variable.

For example, given the regulation statement considered previously the parser will generate the following intermediary form:

IF (A: the occupancy sub-group of the building is A1) AND EITHER (B: the element of structure is a separating wall) OR (C: the height of the building is greater than 28 metres) OR (D: the element of structure is a compartment wall) AND (E: the height of the building is greater than 15 meters) THEN (E: the minimum fire resistance of the element of structure is 60 minutes) ELSE (the fire resistance is 30 minutes)

which will then be mapped to:

IF A AND OR (B OR (C (D AND E))) THEN X ELSE Y

With this approach the user of the system has to provide data using text strings that exactly match those used in the rules. This is because, as mentioned earlier, text strings between logical connectives are treated as constants. In practice this may be a serious limitation because regulation texts are likely to express the same concept in different ways. Another difficulty with text formalisation is ambiguity of interpretation. For any given formalisation there is no way of determining if the logical structure derived is the one intended by the original author. As mentioned in section 3.4.1 the regulation statement used in this example could be parsed in a different way.

6. Author Support Systems

Most of the work described so far has focused on the published texts of regulatory information and the problems of representing of their content in various ways in computer systems. Some researchers (Garzotto and Paoloni 1989; Stone and Tweed 1991) have turned their attention to the production and maintenance of regulatory information and the kind of system functionality necessary to support authors and administrators.

They argue that the published texts are only the most tangible manifestation of a much more extensive information domain and that the conventional view of Standards as comprising solely the published texts is the source of many of the problems of the production of the information and the basis of many of the difficulties users experience in interpreting and applying the requirements in practice. The drafting of regulatory information requires considerable expertise and necessitates an understanding of the enabling principles of legislation, a knowledge of the underlying intent of a particular Regulation or Standard, a detailed knowledge of individual requirements and a familiarity with exemplars and case histories. This knowledge may be well documented but dispersed and held as minutes and records of meetings, notes of discussions, formal records of judicial determinations, written answers to queries and correspondence, research reports and so forth or it may simply be anecdotal. Partly because of the limitations of conventional documentation and partly because of the statutory status of regulatory information there is a dislocation between this experiential, operational knowledge, necessary for the proper understanding and development of Standards, and the formal knowledge contained in the texts. The objective of the research work on authoring systems cited above is to find ways of binding these two domains of knowledge in order to retain and make explicit expertise in the Standards domain and to ensure the proper maintenance of the domain's information.

It will be evident that the amount of information involved is potentially very large. It is for this reason that the research work cited is less concerned with the representation of the content of documents and more with the problems of information organisation and information retrieval.

A major difficulty of organisation is the problem of recording and maintaining links between semantically related material. The conventional way of retaining this kind of control of the domain information relies on the use of physical filing systems and individual's memories to maintain the appropriate connections and relationships. However, physical file systems become unwieldy, individual's memories are unreliable or there may be changes in personnel. The collective and individual memory that links information and gives it meaning is, then, only too easily diminished or lost. The problems of information retrieval parallel and are linked to the problems of organisation. For any given task, a user typically needs to abstract only a subset of the material available but needs to be assured that all the relevant information has been located. This problem is particularly critical for authors when drafting or revising requirements. Before proposing changes they need to explore the consequences of change relative to existing material and ensure that their proposals reflect and properly respond to current thinking and do not conflict with established principles.

6.1 Hypertext and Expertext

This perceived need to improve information organisation and retrieval for authors has led researchers to adopt hypertext as a primary representational paradigm. However, hypertext is not without its drawbacks. Although it does provide for more convenient and structured access to texts than linear documents, it provides no direct functionality with regard to the content of the text nodes and a proliferation of links has been shown to lead to difficulties of navigation. Research is focusing, therefore, on ways of enhancing and extending the basic hypertext model. Of particular interest are the proposals of Rada and Barlow (1989), Diaper and Rada (1989) and Diaper and Beer (1990). Their proposals seek to combine the properties of both expert systems and hypertext and they have coined the term expertext to describe such a system. They argue that both expert systems and hypertext share the same underlying graph-theoretic model. However, they suggest that there are substantial differences in the two approaches in their treatment of nodes and links. Whereas the nodes in hypertext are semantically rich, in that they contain natural language texts, the links have little or no semantic content. Conversely, whereas the nodes in an expert system are semantically impoverished, because they contain a formalisation of the source information, the links are well specified, typically as predicate names in a rule-based system. Expertext represents an attempt to combine the best properties of expert systems and hypertext and provide systems which have the semantically rich nodes of hypertext and the well specified, computable links of expert systems.

It is interesting to look at this proposal from the perspective of how intelligence is distributed between the user and the system. In an expert system the intelligence lies primarily with the system; the user is usually relegated to responding to system generated queries. In hypertext, by contrast, the intelligence lies largely with the user, who interprets the node content but must also specify the order of link traversal. In expertext there is potentially a more balanced distribution of intelligence. Understanding and interpreting the content of nodes is the user's responsibility and is the task most appropriate to human intelligence. On the other hand, because in the expertext model there exist defined and computable links between nodes, the system's intelligence can be focused on the task of navigation, that is on selecting and presenting the most appropriate set of nodes for the user's consideration. It is argued that this distribution of intelligence should reduce if not eliminate the navigation problems conventionally associated with hypertext.

The expertext model potentially fits well with the problems identified in authoring and maintaining regulatory information. Firstly, the problems of the organisation of material and the maintenance of links across domain information can be resolved by expertext's underlying node-and-link structure. Secondly, the problems of information retrieval can be resolved by expertext's intelligent navigation functionality. Thirdly, expertext is not inherently limited by the problems of scale or change in the domain's information. And finally, it seems possible that the functionality and intelligence of the system can be incrementally enhanced as experience in its use is gained.

A particular architecture for expertext was proposed by Rada and Barlow (1989). This has subsequently been developed and termed Headed Record Expertext (HRExpertext) by Diaper and Rada (1989) and Diaper and Beer (1990). The underlying model of HRExpertext remains the node-and-link semantic net. Nodes, however, consist of headed records which have two parts. These are, a record containing the user readable, natural language texts (or potentially any form of user understandable material) and a header which contains an abstraction of the semantic content of the record. Header material is a formalisation of the record's content and is, therefore, an impoverished representation of the record but it has the property that it can be used computationally by the system to select and make available the records consistent with a user's objectives.

6.2 Argumentation

We have noted that a major objective of current work is to bind the experiential and operational information generated by the administration and application of standards to the document domain. The non-document domain of information we may term argumentation. A continuing research problem is to find an appropriate rhetoric model for organising and structuring this kind of information and linking it to the documentation domain. What is required is a method of organising texts which allows a user to expose key relationships within the argumentation material and to make explicit the inherent processes of dialogue and negotiation. Two candidate models which have been proposed are Issue Based Information Systems (IBIS) (Kunz and Rittel 1970; Conklin and Bergeman 1988) and Rhetorical Structure Theory (RST) (Mann and Thomson 1987; Mann et al. 1989).

The IBIS model suggests that argumentation material can be characterised as being the end-product of a process of dialogue or debate, involving a number of participants, about one or more issues. An issue in this sense is simply a question or unresolved problem. IBIS provides a framework for structuring and recording the elements of information generated by this process of issue resolution in a node-and-link representation. IBIS offers a relatively restricted set of node and link types. For example, nodes may be issues, positions or arguments where positions *respond_to* issues and arguments *support* or *object_to* a position.

RST, by contrast, is not primarily concerned with dialogue but with providing a way of analysing and structuring texts as written monologue . It works by splitting the text into fragments of any length and then linking these fragments to form a hierarchy. The links between fragments are commonly of a nucleus-satellite form, in which one node is ancillary to the other. An important difference between the RST and IBIS approach is that nodes in RST can be groups of sub-nodes. This allows links to be placed between large sections of the text, each of which may themselves contain a network of links and sub-nodes. RST provides a more sophisticated representation than IBIS, both in terms of the number of different relations and the definition of restrictions on how the relations should be applied. Typical relation types in an RST graph might be *background* , *concession* , *contrast* , *elaboration*, *motivation* or *means*. IBIS and RST are to some extent complimentary and some hybrid model might be desirable. For example, it may be possible to merge the different types of relations to provide a single set of relations with a rich variety of types; or to use IBIS to map out the dialogue structure of the argumentation material and then use RST relations to map out the fine detail of the issues, arguments and so forth. However, it seems likely that there is no abstract way of resolving the issue of modelling the information in the argumentation domain and that it can only be resolved by pragmatic experimentation in an application context.

7. Conclusion

The paper has reviewed the wide variety of approaches to using information technology which are being experimented with to provide support for users and authors of regulatory information. Practical success to date has been mainly limited to systems which focus on information retrieval functions. The more advanced technologies, such as expert systems, are appealing but seem largely restricted to small scale applications or prototypes. This type of technology, which attempts to represent formally the content of regulatory texts, typically entails one or more transformations or mappings of the source texts. At the moment, these transformation processes are not automated and involve difficult technical, conceptual and practical issues, not least of which are problems of scale and change in the domain information and problems of ambiguity and the preservation of meaning. It is recognised (Reed 1991) that, whilst work on the representation of the content of regulatory documents is technically challenging and of research interest, practical progress will depend upon establishing a common and more rigorous development framework and ultimately on its acceptance by Standards' agencies and the industries involved.

Acknowledgements

The first author is grateful to British Gas and The Fellowship of Engineering for financial support through a Senior Research Fellowship.

References

Bourdeau, M. (1991) The CD-REEF: The French Building Technical Rule on CD-ROM. In Computers and Building Regulations (Eds. K. Kahkonen and B. Bjork), VTT Symposium 125, Technical Research Centre of Finland.

Brownston, L., R. Farell, E. Kant and N. Martin (1985) Programming Expert Systems in OPS5. Addison Wesley.

Clark, K. (1987) Negation as Failure. In Logic and Data Bases (Eds. H. Gallaire and J. Minker), Plenum, New York, 1978.

Clocksin, W. and C. Mellish (1984) Programming in Prolog (2nd Edition). Springer-Verlaag.

Conklin, J. and M. Begeman (1988) gIBIS: A Hypertext Tool for Exploratory Policy Discussion. ACM Transactions on Office Information Systems, 6(4); pp. 303-331.

Cornick, S.M., D.A. Leishman and J.R. Thomas (1991) Integrating Building Codes into Design Systems. In Computers and Building Regulations (Eds. K. Kahkonen and B. Bjork), VTT Symposium 125, Technical Research Centre of Finland.

Crawley, D.B. and J.J. Boulin (1991) ENVSTD: A Computer Program for Complying with Building Envelope Requirements. In Computers and Building Regulations (Eds. K. Kahkonen and B. Bjork), VTT Symposium 125, Technical Research Centre of Finland.

Diaper, D. and M. Beer (1990) Headed Record Expertext and Document Applications. In Proceedings of Hypertext Update; pp. 62-69; Unicom Seminars Ltd., London.

Diaper, D. and R. Rada (1989) Expertext : Hyperising Expert Systems and Expertising Hypertext. In Proceedings of Hypermedia/Hypertext and Object Orientated Databases; pp. 124-152; Unicom Seminars Ltd; Brunel University.

Fenves, S.J. (1966) Tabular Decision Logic for Structural Design, Journal of the Structural Division, Vol 92, No. ST6, June, 1966.

Fenves, S.J., R. Wright, F. Stahl and K. Reed (1987) Introduction to SASE: Standards Analysis, Synthesis, and Expression. NBSIR 87-3513, National Bureau of Standards.

Gardner, A. (1987) An Artificial Intelligence Approach to Legal Reasoning, Chapter 2-4, The MIT Press, 1987.

Garrett, J.H., Jr. and S.J. Fenves (1986) Knowledge-Based Standards Processing. Int. Journal of Artificial Intelligence in Engineering, Vol 1(1), pp3-14, 1986.

Garzotto F. and P. Paolini (1989) Expert Dictionaries: Knowledge-based Tools for Explanation and Maintenance of Complex Application Environments. In Proceedings of the Second international Conference on Industrial and Engineering Applications of AI and Expert Systems IEA/AIE-89; ACM Press, pp. 126-134.

Harris, J.R. and R.N. Wright (1980) Organisation of Building Standards: Systematic Techniques for Scope and Arrangement. Building Science Series NBS BSS 136, National Bureau of Standards, Washington, D.C.

Kahkonen, K, B. Bjork and P. Huovila (1991) The Use of Conceptual Data Modelling as a Basis For Regulations Software Development. In Computers and Building Regulations (Eds. K. Kahkonen and B. Bjork), VTT Symposium 125, Technical Research Centre of Finland.

Kunz, W. and H.W.J. Rittel (1970) Issue as Elements of Information Systems. Working Paper 131. Centre for Planning and Development Research, University of California, Berkley, CA, 1970.

Mann, W. C., C.M.I.M. Mathiesson and S.A. Thomson (1989) Rhetorical Structure Theory and Text Analysis. Isi/rr-89-242, Information Sciences Institute, University of Southern California.

Mann, W.C. and S.A. Thompson (1987) Rhetorical Structure Theory: A Framework for the analysis of Texts. Isi/rs-87-185, Information Sciences Institute, University of Southern California.

Rada R. and J. Barlow (1989) Expertext : Expert Systems and Hypertext. In Proceedings of EXSYS'89 Conference, IITT-International, Paris.

Rasdorf, W.J. and T.E. Wang, Generic Design Standards Processing in an Expert System Environment. ASCE Journal of Computing in Civil Engineering, 2(1), 68-87, 1988.

Reed, K.A. (1991) Building Regulations and Standards in the Computer Age. In Computers and Building Regulations (Eds. K. Kahkonen and B. Bjork), VTT Symposium 125, Technical Research Centre of Finland.

Rosenman, M.A. and J.S. Gero (1985) Design Codes as Expert Systems. Computer-Aided Design, Vol 17(9), 1985.

Sergot, M.J., F. Sadri, R.A. Kowalski, F. Kriwaczek, P. Hammond and H.T. Cory (1986) The British Nationality Act as a Logic Program, Communications of the ACM, Vol 29, No 5, May 1986.

Stone D. and D.A. Wilcox (1987) Intelligent Systems for the Formulation of Building Regulations. In Proceedings of 4th International Symposium on Robotics and Artificial Intelligence in Building Construction, Haifa, Israel, June 1987.

Stone, D. and D.A. Wilcox (1988) Intelligent Information Systems for Building Standards. In Proceedings of EuropIA 88, Applications of Artificial Intelligence to Building, Architecture and Civil Engineering, Paris.

Stone, D. and C. Tweed (1991) Representation Issues in the Design of An Intelligent Information System for Building Standards. In Building Design. In Computers and Building Regulations (Eds. K. Kahkonen and B. Bjork), VTT Symposium 125, Technical Research Centre of Finland.

Susskind, R. (1987) Expert Systems in Law, Part 2-3, Oxford University Press.

Topping, B. and K. Kumar (1989) Knowledge Representation and Processing for Structural Engineeering Design Codes. Eng. Appli. of AI, Vol 2, September 1989.

Tuominen, P. (1991) Fire Expert - An Expert System for Fire Safety Regulations. In Building Design. In Computers and Building Regulations (Eds. K. Kahkonen and B. Bjork), VTT Symposium 125, Technical Research Centre of Finland.

Vanier, D. (1991) A Parsimonious Classification System to Extract Project-Specific Building Codes. In Computers and Building Regulations (Eds. K. Kahkonen and B. Bjork), VTT Symposium 125, Technical Research Centre of Finland.